RESEARCH IN MARITIME HISTORY
NO. 32

TANKERS IN TROUBLE: NORWEGIAN SHIPPING AND THE CRISIS OF THE 1970S AND 1980S

Stig Tenold

International Maritime Economic History Association

St. John's, Newfoundland
2006

ISSN 1188-3928
ISBN 0-9738934-2-7

Research in Maritime History is available free of charge to members of the International Maritime Economic History Association. The price to others is US$25 per copy, plus US$5 postage and handling.

Back issues of *Research in Maritime History* are available:

No. 1 (1991)   David M. Williams and Andrew P. White (comps.), *A Select Bibliography of British and Irish University Theses about Maritime History, 1792-1990*

No. 2 (1992)   Lewis R. Fischer (ed.), *From Wheel House to Counting House: Essays in Maritime Business History in Honour of Professor Peter Neville Davies*

No. 3 (1992)   Lewis R. Fischer and Walter Minchinton (eds.), *People of the Northern Seas*

No. 4 (1993)   Simon Ville (ed.), *Shipbuilding in the United Kingdom in the Nineteenth Century: A Regional Approach*

No. 5 (1993)   Peter N. Davies (ed.), *The Diary of John Holt*

No. 6 (1994)   Simon P. Ville and David M. Williams (eds.), *Management, Finance and Industrial Relations in Maritime Industries: Essays in International Maritime and Business History*

No. 7 (1994)   Lewis R. Fischer (ed.), *The Market for Seamen in the Age of Sail*

No. 8 (1995)   Gordon Read and Michael Stammers (comps.), *Guide to the Records of Merseyside Maritime Museum, Volume 1*

No. 9 (1995)   Frank Broeze (ed.), *Maritime History at the Crossroads: A Critical Review of Recent Historiography*

No. 10 (1996)   Nancy Redmayne Ross (ed.), *The Diary of a Maritimer, 1816-1901: The Life and Times of Joseph Salter*

No. 11 (1997)   Faye Margaret Kert, *Prize and Prejudice: Privateering and Naval Prize in Atlantic Canada in the War of 1812*

No. 12 (1997)   Malcolm Tull, *A Community Enterprise: The History of the Port of Fremantle, 1897 to 1997*

No. 13 (1997)   Paul C. van Royen, Jaap R. Bruijn and Jan Lucassen, *'Those Emblems of Hell'? European Sailors and the Maritime Labour Market, 1570-1870*

No. 14 (1998)   David J. Starkey and Gelina Harlaftis (eds.), *Global Markets: The Internationalization of The Sea Transport Industries Since 1850*

No. 15 (1998)	Olaf Uwe Janzen (ed.), *Merchant Organization and Maritime Trade in the North Atlantic, 1660-1815*
No. 16 (1999)	Lewis R. Fischer and Adrian Jarvis (eds.), *Harbours and Havens: Essays in Port History in Honour of Gordon Jackson*
No. 17 (1999)	Dawn Littler, *Guide to the Records of Merseyside Maritime Museum, Volume 2*
No. 18 (2000)	Lars U. Scholl (comp.), *Merchants and Mariners: Selected Maritime Writings of David M. Williams*
No. 19 (2000)	Peter N. Davies, *The Trade Makers: Elder Dempster in West Africa, 1852-1972, 1973-1989*
No. 20 (2001)	Anthony B. Dickinson and Chesley W. Sanger, *Norwegian Whaling in Newfoundland: The Aquaforte Station and the Ellefsen Family, 1902-1908*
No. 21 (2001)	Poul Holm, Tim D. Smith and David J. Starkey (eds.), *The Exploited Seas: New Directions for Marine Environmental History*
No. 22 (2002)	Gordon Boyce and Richard Gorski (eds.), *Resources and Infrastructures in the Maritime Economy, 1500-2000*
No. 23 (2002)	Frank Broeze, *The Globalisation of the Oceans: Containerisation from the 1950s to the Present*
No. 24 (2003)	Robin Craig, *British Tramp Shipping, 1750-1914*
No. 25 (2003)	James Reveley, *Registering Interest: Waterfront Labour Relations in New Zealand, 1953 to 2000*
No. 26 (2003)	Adrian Jarvis, *In Troubled Times: The Port of Liverpool, 1905-1938*
No. 27 (2004)	Lars U. Scholl and Merja-Liisa Hinkkanen (comps.), *Sail and Steam: Selected Maritime Writings of Yrjö Kaukiainen*
No. 28 (2004)	Gelina Harlaftis and Carmel Vassallo (eds.), *New Directions in Mediterranean Maritime History*
No. 29 (2005)	Gordon Jackson, *The British Whaling Trade*
No. 30 (2005)	Lewis Johnman and Hugh Murphy, *Scott Lithgow: Déjà vu All Over Again! The Rise and Fall of a Shipbuilding Company*
No. 31 (2006)	David Gleicher, *The Rescue of the Third Class on the Titanic: A Revisionist History*

Research in Maritime History would like to thank Memorial University of Newfoundland for its generous financial assistance in support of this volume.

Table of Contents

Tables in the Text

List of Illustrations

Preface

I first began to analyse the shipping crisis of the 1970s and the Norwegian predicament in a Master's thesis in economic history at the Norwegian School of Economics and Business Administration in the mid-1990s. As often happens, this early work generated more questions than answers.

There were two reasons for my interest in modern maritime history. First, as a boy growing up in Norway in the 1980s, shipping was very much a part of my daily life and cultural heritage. I lived some 300 metres from the headquarters of Hilmar Reksten, who – in his own words – was the world's largest independent tanker owner. Indeed, I delivered newspapers to his shipping company every morning. At the time, I did not know that the decreasing number of subscriptions reflected the decline of Reksten's empire. But as an economics student I subsequently learned about the principle of comparative advantage. The shipping sector is perhaps the best arena in economic history for the analysis of questions which relate to both the international and the Norwegian economies.

The second reason for my interest in maritime history is even more straightforward: Professor Helge W. Nordvik. Helge was renowned for his work in maritime history both in Norway and around the world, and his enthusiasm for shipping and shipping history was contagious. When I began my doctoral studies I asked Helge to be my supervisor, and many of the questions, hypotheses and approaches were formulated in close cooperation with him. Helge died suddenly in October 1998. My thesis, "The Shipping Crisis of the 1970s: Causes, Effects and Implications for Norwegian Shipping," was defended in 2001. While I hope that it met Helge's high standards, it certainly was guided by one of his favourite mantras: "If it's worth doing, it's worth doing well."

This book is a significantly restructured and updated version of the thesis. While the main arguments are unchanged, the structure of the analysis is new. In particular, I have tried to write the text so that the chapters can be read more or less independently. Consequently, someone with a particular interest in business history may go directly to chapter eight, while those with a strong dislike for statistical excursions can turn a blind eye to chapter seven without losing the thread of the argument.

I have presented the main findings in various academic settings and to a number of regional shipowners' organisations in Norway. By discussing the topic with shipping industry insiders – some of whom were among the biggest victims of the crisis – I gained new insights. Not surprisingly, a number of industry participants claimed that they had warned of an imminent crisis shortly before the freight market collapsed. While I do not wish to cast aspersions on their recollections, it is significant that not a single one of these prophets has so far managed to substantiate such claims with documents or other hard evidence.

I am also greatly indebted to Professor Lewis R. "Skip" Fischer. After Helge passed away, he gave fruitful advice on the direction of my thesis. More important, however, he took over the "mentoring" responsibility that supervisors sometimes take. I am extremely grateful for his help and encouragement. I would also like to thank the International Maritime Economic History Association for giving me the opportunity to publish this book in its *Research in Maritime History* series.

This book is dedicated to the memory of Helge W. Nordvik

Stig Tenold
Bergen, Norway
October 2006

Chapter 1
Introduction

How bad might a "worst case" scenario be? There are some examples in economic history of rapidly disappearing markets, usually in connection with conflicts, the introduction of new products or revolutionary technologies. Yet the shipping crisis of the 1970s and 1980s illustrates that even without such disruptions, sustained and rapid expansion might abruptly be replaced by decline.

In the marine sector the starting point was favourable. By 1973 the market for oil transports had grown more or less without interruption for a century. Indeed, in the previous decade annual growth in tanker transport demand had averaged almost seventeen percent. Most shipowners acted accordingly, ordering new capacity on the belief that strong growth would continue. Although a limited number of pessimists predicted lower growth rates, no one expected an absolute decline. A decade later the situation was quite different. By 1983 the demand for tanker transport had more than halved compared with the peak, and more than fifty percent of the tanker fleet was superfluous. Not even the most pessimistic observer would have predicted such a shift.

This book analyses the causes and effects of the most serious crisis in the shipping market in the twentieth century. The analysis is conducted at three progressively narrower levels. Initially, the basis for the crisis at the macro level is examined. The starting point is the tanker sector, where the problems surfaced first and were most severe. After a discussion of the effects and spread of the crisis, we move to the national level. The initial focus here is Norwegian shipping and shipowners in general. Finally, we take the analysis to an even more detailed level, using a business history perspective to focus on the strategies and fates of four individual tanker owners.

In some ways, the structural shift in the shipping market after 1973 mirrors the general development of the world economy, although the fluctuations within shipping were more pronounced. This was the case in the period before 1973, when shipping demand grew faster than world trade in general. It was also very clear after 1973: whereas most sectors of the economy suffered from lower growth rates, there was an absolute decline in demand within a number of sectors in the shipping industry.

The shipowners' reaction to the beneficial conditions before the onset of the shipping crisis is important to explain developments after the market collapsed. The book therefore begins with a brief presentation of the pre-crisis experience. The rapid demand growth before 1973 played an important role in shaping the beliefs and strategies of shipowners and other participants in the shipping market.

Two lessons from this period form a crucial part of the subsequent analysis. First, the stable demand growth removed some of the apparent risks of shipping investments – few shipowners were unable to take advantage of the beneficial conditions. In most segments, demand grew substantially from one year to the next. The basis in part was increasing volumes, but structural changes, such as increases in average voyage length which increased the need for carrying capacity, also played a role. The rapid expansion of demand fuelled shipowners' optimism – both the size of the world fleet and the average size of the ships increased substantially. Second – and closely related to the high demand growth – shipowners who were willing to invest in new tonnage were generally amply rewarded. Although the shipping market was characterised by its familiar cycles, the good years occurred sufficiently often to deliver positive – sometimes massive – operating profits. Moreover, second-hand tonnage did not decline much in value, giving an additional bonus to shipping companies that were able to time their sales and purchases correctly. The relatively stable vessel values might sound slightly paradoxical given the rapid technological development of shipping in this period. During most of the 1960s, however, there was a pent-up demand for new shipping capacity. The shipbuilding industry was unable to deliver the volumes that shipowners wanted, and the size of the order books increased.

The optimism – and investments – that the experiences of the first postwar decades brought about came back to haunt shipping for a considerable time after 1973. The tanker sector, which had been the most vibrant segment of the industry before the crisis, plunged to historically low depths. In chapter 3, a simple supply-and-demand framework is used to analyse the causes of this problem. On the demand side, the effects of increases in oil prices of course had important explanatory power. Yet the analysis also demonstrates that spatial shifts in oil production and changes in the chartering policies of oil companies helped to spread and prolong the crisis.

Nevertheless, looking only at demand ignores an important side of the story; we also need to understand why the fleet grew despite the fact that the market was characterised by overcapacity. The explanation of the growth of tanker supply in a depressed market is multifaceted. Shipowners' initial optimism – and their record amounts of tonnage on order when the crisis erupted – is one important factor. Moreover, their failure to understand the severity of the crisis – initially treating it as a business cycle phenomenon rather than as a structural problem – also contributed to worsening the situation and lengthening the crisis. Nevertheless, despite being the ones who made the investments, shipowners should not shoulder all the blame for the strong growth of transport capacity. The role of a number of other agents also needs to be considered. The analysis of the causes of the tanker crisis therefore takes into account the ways in which shipbuilders, financial institutions and governments contributed to the oversupply.

The actions of these agents are also important in explaining why the tanker crisis turned into a general shipping crisis, which is one of the topics in chapter 4. With combination carriers constituting a direct link between tanker shipping and other segments, the contagion was relatively rapid. In the longer run, other factors aggravated the situation. The conversion of tanker new buildings to other types of vessels amplified the overcapacity outside the tanker sector. Moreover, faced with declining activity, shipyards – usually funded through government subsidies – managed to secure new orders despite a looming surplus of dry bulk and general cargo ships. Coupled with an international recession, the results were disastrous. From the second half of the 1970s onwards, practically all segments of the shipping market were depressed. Shipping companies were hurt as freight rates, profits and vessel values dwindled.

The first stage of the shipping crisis was caused by the growing supply of transport capacity and demand that failed to live up to the shipowners' expectations. In chapter 5 the second stage of the crisis is discussed. Although the tanker fleet was in fact declining from 1978 onwards, an absolute reduction in oil transport demand meant that the market for large tankers still was far from equilibrium. Indeed, the balance between supply and demand was at its worst in April 1983, almost ten years after the Organization of Petroleum Exporting Countries (OPEC) had first raised oil prices. Although slow steaming, the acceptance of part cargoes and the use of tankers for storage purposes absorbed some of the surplus, more than fifty million deadweight tons (dwt) of tankers had been mothballed.

Periods of depression often transform industries. Parallel with the shipping crisis, there were significant shifts in the registration and ownership of the world fleet. Two groups of up-and-coming registries – Asian countries and the so-called flags of convenience (FOCs) – increased their presence in international shipping substantially. The basis for the growth of these countries, as well as the reduced importance of the Traditional Maritime Nations (TMNs), is discussed within the framework of the shipping crisis. In particular, we will see that the same competitive forces that led to the relocation of manufacturing from countries in the Organisation for Economic Co-operation and Development (OECD) to Asia were also important in explaining the end of their maritime hegemony. The exploration of the shifts in world registry is the final part of the book dealing with the international level, and it paves the way for an examination of Norway, one of the best examples of the misfortunes and decline of the TMNs.

In the early 1970s Norway, with approximately 0.1 percent of the world's population, was the world's fourth greatest maritime nation, and its tonnage comprised approximately ten percent of the world fleet. The importance of Norwegian shipping in an international context was mirrored by its significance to the Norwegian economy. Ever since the days of sail in the nineteenth century, income from the export of shipping services had comprised between a third and half of Norwegian export revenues. In the first postwar decades – with limited possibilities for cross-border investments – large financial resources were channelled into shipping. Accordingly, Norway not only had a large fleet in international terms but also one that consisted primarily of modern and technologically-advanced vessels.

The basis for the crisis in Norwegian shipping is analysed in chapter 6. The Norwegians had been well situated to compete in an era of growth, but when the market collapsed they were more severely affected than their competitors. By 1987 the volume of Norway's merchant marine had been reduced by approximately eighty percent relative to the peak ten years earlier. Three salient features that differentiated the Norwegians from shipowners in other countries are evaluated in order to explain the disparate development. One element was fleet structure, where the Norwegian emphasis on mammoth tankers and bulk carriers is explained in terms of comparative advantages in the prices of labour and capital. The second element was Norwegian chartering policy, based upon the oft-cited claim that Norwegian owners were risk-takers, generally operating their ships on a voyage-by-voyage basis. Surprisingly, a closer analysis of new sources shows that this claim is incorrect, and one of the most common explanations of the Norwegian shipowners' predicament is thus mainly refuted. The final feature explaining the severity of the crisis in Norwegian shipping was the level and types of new building contracts. While it is shown that the type and size of ships on order was not significantly different than their foreign competitors, the sheer volumes presented a problem for the Norwegians.

Just as the shipping crisis led to structural changes in the world fleet, there was also a transformation of Norwegian shipping. Chapter 7 is an empirically-based analysis of the shifts in Norwegian shipping, centring on tonnage, the number of companies and their spatial dispersion. The crisis coincided with increased concentration, both with regard to shipping companies and the spatial distribution of ownership. These transformations are analysed using a purpose-built database, which shows that the number of Norwegian shipping companies declined by approximately two-thirds during the crisis. The analysis also demonstrates that the size of a company was important in determining its fate during the first part of the crisis; those who were forced to ship the oars were mainly smaller businesses. But after 1978 companies at all levels in the Norwegian shipping hierarchy were beset by the problems, despite efforts by the authorities to help. There also seems to have been significant differences in the extent to which shipowners in different parts of the country managed to cope with the crisis.

Most shipping statistics use dwt or gross registered tonnage (grt) to measure fleet size. While this would be sufficient if all ships were the same, it actually cloaks a number of important properties of the ships. At the end of chapter 7 an alternative approach is used in which the conventional tonnage figures are adjusted for the type and sophistication of the ships. Based upon compensated gross register tons (cgrt), the decline of the Norwegian fleet becomes less significant. This new measure, which has previously been used primarily in shipbuilding, illustrates the structural shift in the Norwegian fleet during the crisis. The mammoth tankers and bulk carriers were replaced by more sophisticated, but smaller, vessels which were more valuable on a per-ton basis.

The analysis of Norwegian shipping gives us some indications of the reasons for the fortunes and misfortunes of shipping companies in general. But since a more thorough examination of the practices of different shipowners may reveal the challenges that confronted them, in chapter 8 we will look at the experiences of four tanker owners. The analysis illustrates how the fates of their companies were closely intertwined with strategic decisions taken shortly before and after the onset of the crisis. The varying fortunes illustrate how small nuances in apparently similar business strategies may make the difference between survival and bankruptcy. Shipping is a business in which vast fortunes are made rapidly, and these rags-to-riches stories are generally well-known. Nonetheless, the four case studies illustrate that one or two bad decisions can drain a company's resources just as quickly.

The aim of this book is to add to our knowledge of recent maritime history by examining the transformation of the industry during a period of rapid change. It may seem paradoxical that one of the most important lessons from this study is that history can be deceiving; uncritical extrapolation of past trends became the bane of many shipowners, in Norway and elsewhere, in the 1970s and 1980s.

Chapter 2
The Shipping Market after World War II

Shipping and shipbuilding underwent a series of major changes in the first decades after World War II. Growth in world trade, with corresponding increases in the demand for shipping services and technological improvements, laid the foundation for a drastic transformation of the shipping market. The movement towards increased globalisation that began in the nineteenth century was interrupted in the interwar period. Seaborne trade increased only slightly, and it was primarily the rapid growth in the petroleum trade that prevented an overall reduction. During most of the 1930s the world fleet declined due to the scrapping of obsolete tonnage and limited reinvestment. But growth resumed after 1945, and technological innovations in shipping and shipbuilding enabled shipowners to benefit from the improved conditions.

Strong growth and structural changes paved the way for the transformation of the world fleet. Both its structure and the vessels themselves reflected growth and new opportunities. One prominent shift was the increasing significance of the oil trade, with liquid bulk transports becoming the dominant segment of the shipping industry. Moreover, the period from 1950 to the early 1970s was marked by the first stage of the containerisation of the liner trades and the breakthrough of bulk shipping into a number of new cargoes. The result was increased specialisation and reduced linkages between the various shipping sectors. The maritime economist Martin Stopford has written that:

> [i]n 1945 the world merchant fleet consisted of passenger ships, liners, tramps and a small number of tankers. Few vessels used for cargo transport were larger than 20,000 dwt. By 1975 the fleet had changed out of all recognition and all the major trades had been taken over by specialized ships.[1]

The demand for transport services is closely related to the level of world trade. The first quarter-century after the end of the World War II was characterised by strong growth in international trade.[2] Economic growth was

---

[1]Martin Stopford, *Maritime Economics* (London, 1988; 2nd ed., London, 1997), 57.

[2]The demand for transport is determined by the volume rather than the value of world trade. Moreover, the share of seaborne transport in total transport is not constant, and changes in average distances may also affect the demand for tonnage.

strong in most Western nations in the initial decades after 1945. Yet a characteristic feature of international economic development was that world trade grew even faster than production. While global production trebled from 1953 to 1973, the volume of international trade grew by 350 percent.[3] And the volume of goods transported by sea increased even more rapidly, swelling from 490 million to over 3200 million tons, or by more than 550 percent, between 1948 and 1973 (see figure 2.1). These increasing volumes were augmented by a rise in average shipping distances, particularly in the latter part of the period.

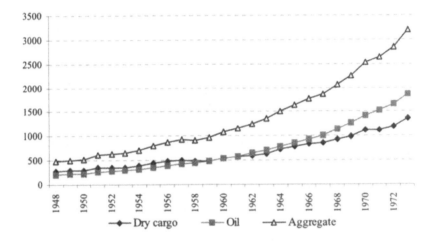

Figure 2.1:   Development of World Seaborne Trade, 1948-1973 (million tons)

Source: Organisation for European Economic Cooperation/Organisation for Economic Co-operation and Development (OEEC/OECD), *Maritime Transport*, as cited in Gelina Harlaftis, *A History of Greek-Owned Shipping: The Making of an International Tramp Fleet, 1830 to the Present Day* (London, 1996), 250. The OEEC/OECD claims that the figures originated with the UN, but it uses figures from Fearnley and Egers Chartering Co. Ltd., *Review* (Oslo, various years) for the later period. One problem is the inconsistency between the Fearnley and Eger figures and those of the OECD in the years in which both present data. OECD figures tend to overestimate, and Fearnley and Eger's tend to underestimate, oil transport in the early 1960s. The trend is nevertheless clear, although the point at which oil exceeds other cargoes is shifted from 1960 to 1968 when Fearnley and Eger's data are used.

---

[3]Herman van der Wee, *Prosperity and Upheaval. The World Economy 1945-1980* (London, 1986), 260. The volume of manufacturing exports increased by more than 500 percent.

Despite the strong underlying increase in international trade, there have been substantial variations in the growth rate for different periods and goods. One important feature was the substantial rise in liquid bulk shipping. At the beginning of the 1960s the volume of liquid bulk for the first time exceeded dry bulk. This contributed to – and was a result of – the most important supply-side structural change: the massive growth of the world tanker fleet.

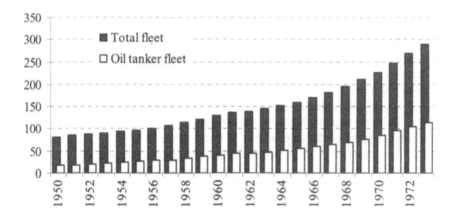

Figure 2.2: World Fleet and the Oil Tanker Fleet, 1950-1973 (million gross registered tons [grt])

Note: Ships over 100 grt only. Combination carriers are not included in the tanker totals.

Source: OEEC/OECD, *Maritime Transport*, as cited in Harlaftis, *History of Greek-Owned Shipping*, 256.

Figure 2.2 shows the strong growth of the world fleet in general and the tanker fleet in particular. But the data on which this is based are conservative; similar analyses, based on deadweight tonnage (dwt) and with a higher minimum size limit, suggest that ships with tanker transport capacity in fact surpassed the non-tanker fleet by the beginning of 1971.[4] Regardless of the method of measurement, it is clear that growing demand enticed shipowners to order new tonnage. At the same time, the supply chain was improved through sharp increases in shipbuilding capacity: tonnage launched increased from 2.1

---

[4]Based upon a lower boundary of 10,000 dwt for tankers, combination carriers and dry bulk vessels, and 1000 grt for other ships; see Fearnley and Egers Chartering Co. Ltd., *Review 1977* (Oslo, 1978), 15.

million grt in 1946 to more than thirty-five million grt by the mid-1970s, four times as much as was launched in 1960.[5]

Paralleling this growth in capacity was a significant shift in the spatial distribution of world shipbuilding. While in 1974 the order books of British yards were three times the level of 1946, Japanese yards had contracts that were twenty-seven times larger. Nonetheless, the fact that the world order book increased every year after 1963 implies that despite a huge increase in capacity shipbuilders were unable to cope with the growing demand.

Victor Norman has claimed that there was a "latent demand surplus" in shipping because demand increased more than expected for most of the postwar period.[6] Even though market conditions varied by time and segment, there were periods that were clearly booms for shipping as a whole. Between 1945 and 1970 there were three such eras with particularly high freight rates.

As international economic relations returned to normal after World War II, there was an immense increase in the demand for shipping. Yet the limited capacity of the shipbuilding industry made it difficult to meet the need. The first boom began in the summer of 1950, related partly to the onset of the Korean War. Stockpiling in Western countries led to short-term demand growth, and freight rates increased rapidly. The boom lasted approximately sixteen months. Due to restrictions imposed on contracting, Norwegian shipowners were unable to take full advantage of these beneficial conditions. Moreover, the favourable contracts that Norwegians were prohibited from honouring due to the ban were assumed largely by Greek shipowners in exile, thus facilitating the expansion of their tanker fleets.[7]

The second boom was launched in the mid-1950s. International political developments again played a leading role. The strong growth of world trade led to an increase in freight rates, a situation that was exacerbated by the closure of the Suez Canal in November 1956. As a result, crude oil transports were redirected around the Cape of Good Hope, which in turn led to a sharp

---

[5]Anthony Slaven, "Marketing Opportunities and Marketing Practices: The Eclipse of British Shipbuilding, 1957-1976," in Lewis R. Fischer (ed.), *From Wheel House to Counting House: Essays in Maritime Business History in Honour of Professor Peter Neville Davies* (St. John's, 1992), 125-127.

[6]Victor D. Norman, "Har det vært for lett å drive internasjonal skipsfart i Norge?" ["Has Shipping Been Too Easy an Enterprise in Norway?"] *Internasjonal Politikk*, No. 1B (Oslo, 1979), 176-177.

[7]Atle Thowsen, "Skipsfart og planøkonomi. Kontraherings- og lisensieringspolitikken overfor norsk skipsfart i den første etterkrigstiden (1945-1953)" ["Shipping and the Mixed Economy: The Contracting and Licensing Policies of Norwegian Shipping in the Initial Postwar Period (1945-1953)"], *Sjøfartshistorisk Årbok 1985* (Bergen, 1986), 23. The contracting prohibition ended by the beginning of 1951.

increase in demand for tankers, especially for the largest ones. After about eighteen months, this boom ended due to a four-percent decrease in OECD industrial production; the reopening of the Suez Canal in April 1957; and an increase in the supply of tonnage.[8] As a result of the suddenly adverse conditions, the lay up-rate in the dry bulk market reached a new postwar high.

The last boom spanned the five months after the Six Day War between Egypt and Israel in 1967. Suez was again closed, increasing the shipping distance between the Persian Gulf and Europe by eighty percent. Tanker rates in the spot market – where transport is sold on a per-voyage basis – were four times higher between June and October than in the previous six months.[9] Moreover, in the period 1966-1968 the average distance of international crude oil shipments increased by twenty-five percent.[10] Spill-over effects ensured that profits in other segments increased as well. As combination carriers previously offered in the dry bulk market entered the booming tanker sector, tonnage available for dry bulk cargoes fell and rates increased.

A number of structural shifts in international trade fuelled this strong growth in seaborne transport (see figure 2.3). One important element was the rise of Japan as a major supplier of manufactured goods. The Japanese share of world exports increased from 0.4 percent in 1948 to more than 6.4 percent by 1973, and the fact that the country was dependent on imports for most of its raw materials further boosted shipping demand.[11] Another factor was the increasing dependence on oil as the main source of energy. Due to the geographic concentration of oil reserves, most industrial nations had to import a considerable portion of their energy requirements, and by far the largest share went by sea. During the 1960s, demand for tankers grew faster than for bulk carriers, and tankers became the most important type of shipping. By 1970, the transport of oil and oil products accounted for more than half the ton-miles of world seaborne trade. As a share of shipping demand, the defining moment was reached in the early 1960s (measured by ton-miles). In 1968, one year

---

[8]The increased supply of tonnage was a result of the regular growth of the fleet, as well as increased utilisation of American tonnage reserves; see Johan Seland, *Norsk skipsfart år for år 1946-1976* [*Norwegian Shipping Year by Year, 1946-1976*] (Bergen, 1994), 76-77.

[9]*Ibid.*, 169.

[10]Calculations based on data from Fearnley and Egers Chartering Co. Ltd., *Review 1972* (Oslo, 1973), 8.

[11]Data on the Japanese share of world trade are from World Trade Organisation (WTO), *International Trade Statistics 2005* (Geneva, 2005), table II.2.

after the second postwar closure of the Suez Canal, tankers accounted for approximately sixty percent of all seaborne transport movements.[12]

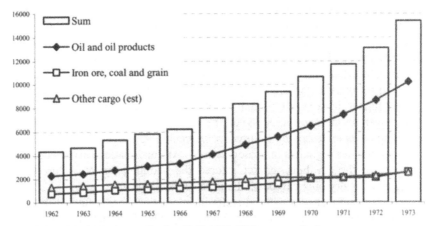

Figure 2.3:   World Seaborne Trade, 1962-1970 (billion ton-miles)

Note: One billion equals one thousand million.

Source: Fearnley and Eger, *Review*, various years.

As a result of growing world trade, shipping transport demand grew considerably in the period after 1945. At the same time, the supply of tonnage increased, not only as a result of an increase in world shipbuilding capacity but also due to technological developments. Technological improvements were evident in a number of areas. One was the large increase in vessel size to take advantage of economies of scale. The closure of the Suez Canal, the growing importance of the Middle East in oil production and Japan's rapidly increasing oil imports were significant factors in the strong growth in average tanker size. In 1950 few tankers exceeded 30,000 dwt, but seven years later, after the first Suez crisis, almost half the tanker tonnage on order consisted of vessels over this threshold.[13] The second closure of the Suez Canal had similar effects: the average size of tankers on order increased from approximately 63,000 dwt in

---

[12]All calculations based on figures from Fearnley and Egers Chartering Co. Ltd., *Review 1972* (Oslo, 1973), tables 1 and 2. The distinction between the OECD and Fearnley and Eger's figures implies that the shift occurred earlier if OECD data are used. The difference between the tankers' share of seaborne trade measured in tons and ton-miles is due to the relatively longer average distance of oil transport.

[13]Gunnar Nerheim and Kristin Øye Gjerde, *Uglandrederiene – verdensvirksomhet med lokale røtter* [*The Ugland Shipowning Companies – International Business with Local Roots*] (Grimstad, 1996), 104.

1965 to more than 160,000 dwt by 1973. The escalation in size of the existing tanker fleet was less pronounced, but it nevertheless doubled, from approximately 28,000 dwt in 1965 to more than 58,000 dwt by 1973.

The movement toward larger ships was also evident in dry bulk shipping. Indeed, the whole concept of dry bulk shipping – the use of specialised bulk carriers as opposed to general cargo vessels – was affected by some of the same factors as tankers. Demand for bulk staples, such as iron ore, bauxite, coal and phosphates, was high due to increased manufacturing output. Moreover, shifts in the international pattern of trade amplified the demand for shipping. According to Anders Martin Fon, "the sustained and rapid growth in Japanese industrial output was the most important factor in explaining why bulk trades increased so dramatically in the 1960s."[14] At the same time, technological improvements, in particular in cargo handling, increased efficiency and enabled the use of ever-larger vessels. The average size of a dry bulk carrier increased by about fifty percent between 1965 and 1973.

Dry bulk vessels – purpose-built for the transport of cargoes such as iron ore and coal – illustrate this increasing specialisation. While general cargo vessels could transport a vast array of goods, the new ships were dedicated to specific cargoes. In addition to conventional tankers and bulk carriers, the growth of gas and chemical tankers, container ships and car carriers underscores the increased fragmentation. The basis for this transformation was twofold. On the one hand, growing trade ensured a "minimum demand" for a number of "new" commodities. Without sufficiently large cargoes, the economic rationale for such specialisation was non-existent. On the other hand, improved cargo handling and other technological improvements gave a boost to the supply side. As vessels became larger, more specialised and more expensive, the time spent in port had to be minimised. Technological improvements facilitated this change.

Liner shipping, which in the 1960s was marked by the dawn of the container era, was one segment in which new technological solutions revolutionised the land-sea interface. Yet according to Frank Broeze, the liner sector initially lagged behind bulk shipping in adopting technological improvements:

> By 1960 often not even fifty percent of ship time was spent at sea, and current technology had reached the limits of its capabilities to achieve improvements in efficiency and productivity. By contrast, in the relatively young bulk sectors of the maritime freight market, ever larger oil tankers and dry-

---

[14]Anders Martin Fon, "En stormakt i tørrbulk. En økonomisk-historisk analyse av norsk tørrbulkfart 1950-1973" ["A Major Dry Bulk Nation: An Economic-Historical Analysis of Norwegian Dry Bulk Shipping 1950-1973"] (unpublished Dr. oecon. thesis, Norwegian School of Economics and Business Administration, 1995), ii.

bulk carriers, operating in synergy with dedicated modern
cargo loading and discharging facilities ashore, indicated the
way ahead towards greater productivity through specialisation and economies of scale.[15]

Containerisation was the way that liners achieved productivity improvements.

Despite advances in vessel technologies and continuous growth in
demand, shipping continued to be characterised by relatively violent business
cycles. By analysing these cycles in the postwar period, Arnljot Strømme
Svendsen has found that thirty-four of the 120 months in the 1950s can be
characterised as booms compared to only five in the 1960s.[16] Stable international trading and monetary conditions contributed to the massive growth of
the global economy in the first decades after World War II. By the end of the
1960s, however, it was evident that the existing monetary regime, characterised by fixed exchange rates and full currency convertibility, no longer reflected the actual balance of power in the world economy.

A liberal commercial and monetary regime was important for the
growth of international trade prior to 1970. Indirectly, they were important to
shipowners because they facilitated the growth in demand for seaborne transport. So too did fixed exchange rates, for they created a stable, consistent and
predictable framework in which it was relatively easy to predict costs and revenues. Since shipping was an international industry, shipowners usually had
income and expenses in different currencies, and exchange rate adjustments
therefore influenced profitability calculations for new building contracts and
charter parties. Differences in the economic development of the Western countries after 1945 led to major disequilibria in the international economy. By the
beginning of the 1970s the conditions for international settlement were radically changed with the collapse of the gold-exchange standard, underpinned by
the US dollar. The 1970s were marked by lower growth rates for manufacturing and international trade. Moreover, unemployment and inflation increased
sharply, as did uncertainty regarding the stability of the international economy.

These fundamental shifts in international economic and monetary relations had significant implications for shipping. On the one hand, reduced
growth in international commerce removed the basis for increasing demand for
shipping transport. On the other, the breakdown of international monetary stability increased uncertainty and had both direct and indirect consequences for
shipowners. The history of international shipping in the 1970s was dominated

---

[15]Frank Broeze, *The Globalisation of the Oceans: Containerisation from the
1950s to the Present* (St. John's, 2002), 9-10.

[16]Arnljot Strømme Svendsen, "Skipsfartskonjunkturene i 1970-årene" ["Shipping Cycles in the 1970s"], *Sjøfartshistorisk Årbok 1978* (Bergen, 1979), 209.

by the transformation of the international economy and the extent to which shipowners were able to cope with the structural changes that took place.

Despite what was to come, the 1970s began on a high note. The remarkably strong growth in shipping demand of the 1960s continued in the first part of the 1970s, despite the changes in international monetary relations. The year 1969 had not been particularly good for the shipping industry: the supply of tonnage had grown faster than transport demand, and freight rates were falling. But freight rates rose after June 1970 as developments in Libya and Saudi Arabia reduced oil shipments.[17] As a result, Europe increasingly had to rely upon the Persian Gulf, and these transports had to go around the Cape because the Suez Canal remained closed. In addition, the continuing growth of the Japanese steel industry had a positive effect on dry bulk demand. But a slowing of international trade growth and a consistently large supply of new tonnage led to a fall in freight rates after approximately one year.

In the summer of 1972, freight rates started to rise again. Despite an increase in international trade, a large amount of tonnage, particularly in the dry bulk sector, remained in lay-up. In the dry bulk market the rate increase was a result of the Japanese seamen's strike and an increase in Soviet grain imports. The real boom, however, took place at the beginning of 1973, when conditions were particularly good for tankers operating in the spot market. This development was tied closely to expectations about growth in American oil imports. During 1973, tanker freight rates continued to increase, and between May and September rates shot up from Worldscale (W) 100 to W 475.[18]

The Yom Kippur War in October 1973 had dramatic consequences. Even though it was over within a month, it led to important changes in the international power structure and the world economy. In the market for large tankers, voyage rates fell from more than W 450 in October to W 55 in November.[19] This signalled the start of the shipping crisis.

---

[17]The pipeline Tapline was damaged and the oil supply from Sidon reduced. The result was an increase in transport distances and seaborne transport. Moreover, the new Libyan dictatorship introduced restrictions on the delivery of oil to Europe.

[18]Worldscale represents the daily revenue for a standard tanker, independent of its voyage. In principle, a tanker operator will therefore be indifferent between two offers of Worldscale 100, although there may be special circumstances which cause him to prefer one offer. For a basic introduction to the principles surrounding the use of Worldscale, see Roy Nersesian, *Ships and Shipping - A Comprehensive Guide* (Tulsa, OK, 1981), 21-22. A rate of W 475 indicates 475 percent of the published rates.

[19]Svendsen, "Skipsfartskonjunkturene," 228, does not indicate the route involved, but most probably it is the Persian Gulf to Europe. Fearnley and Egers Chartering Co. Ltd., *Review 1973* (Oslo, 1974), 13, shows that based upon weekly averages, rates for a 100,000-dwt ship from the Persian Gulf to the UK and Europe peaked at W 342 in September before decreasing to W 134 in November and W 77 the next month.

Chapter 3
The Basis of the Tanker Crisis

When discussing the properties of a crisis – whether in shipping or elsewhere – it is sometimes difficult to distinguish between cause and effect. For instance, if we focus on the aggregate level and define the shipping crisis as a mismatch between demand and supply, plummeting freight rates become an effect. But if we consider the plight of shipowners and regard falling profitability as the crisis, the low level of freight rates was a cause rather than an effect. Consequently, it might be fruitful to discuss what constitutes a crisis.

A traditional definition of a depression or slump is that it is characterised by underutilisation of the factors of production and falling demand. In shipping, underutilisation of the capital stock is common. Unlike manufacturing, the shipping industry cannot easily accumulate inventory for future sale. If supply exceeds demand, this is reflected in lower rates. If rates fall below a certain level, ships have to be laid-up, and laid-up ships imply that the capital stock is not being utilised fully.

Even during the so-called "golden age," there were more than six lean months for every fat one. These slim periods were characterised by a surplus of tonnage relative to demand. Indeed, in the late 1950s between seven and eight percent of the dry bulk fleet was laid-up, a rate twice as high as during the shipping crisis of the 1970s.[1] Oversupply of tonnage is a typical example of the underutilisation of resources. Yet despite the fact that prior to 1973 large numbers of ships had sometimes been idle, only in 1958 was this associated with falling demand.

Crises discourage investment in new capacity, but before the 1970s there was substantial investment in shipping. The world fleet grew annually, and rapid technological development led to substantial improvements in the capital stock. Vessels became larger, more specialised and more sophisticated. The large amounts of new tonnage led to a decline in average age. Favourable demand made new investments attractive despite temporary overcapacity.

Thus, any talk of a crisis in postwar shipping before 1973 is inappropriate. The periods of overcapacity were too short to have had significant effects on long-term profitability or on the structure of the industry. Even those shipowners who were hit hardest by overcapacity, and who had to lay-up a considerable share of their fleets, normally managed to weather the storm.

---

[1]Michael Beenstock and Andreas Vergottis, *Econometric Modelling of World Shipping* (London, 1993), 36.

Moreover, demand grew in the long term. There was room for considerable expansion and modernisation, particularly in the tanker sector. Overall, in most years shipowners were able to earn substantial profits.

In the 1970s and early 1980s, on the other hand, the use of the term can be justified. First, the depression – characterised by low rates, over-capacity and an absolute reduction in demand – lasted sufficiently long to affect the financial position of most owners severely. The real recovery took more than a decade to materialise.[2] Second, it was accompanied by fundamental changes in shipping. The fleets of several important shipping companies and maritime nations dwindled. This was accompanied by bankruptcies, flagging-out, increased government involvement and the collapse of shipbuilding, particularly in Western Europe. The dire conditions were also felt in auxiliary industries, including ship finance, insurance, shipbroking and classification.

The crisis of the 1970s and 1980s was the most severe depression in shipping in the twentieth century. Relative to other episodes of overcapacity, the crisis differed in breadth, depth and length. Breadth refers to the number of actors affected; depth is a measure of the seriousness of the conditions; while length denotes the time elapsed before recovery. In all three dimensions, the problems were much larger than anything experienced previously.

To analyse the causes of the shipping crisis, we will deal with demand and supply separately. The aim is to show how historical development influenced the expectations and actions of economic actors. Historical circumstances were necessary conditions for the crisis, the impulses that triggered responses that led ultimately to a severe depression.

In trying to identify a "scapegoat" for the crisis (and for the general recession of the 1970s), it is common to point the finger at the Organization of Petroleum Exporting Countries (OPEC), which played a crucial role in shaping the demand for tankers. In 1973, the Norwegian government's long-term prognosis was that a sharp increase in demand for petroleum and associated products was likely unless there was an major disruption in price or supply.[3] Unfortunately, OPEC's decision to increase the price of oil (see figure 3.1), and the economically- and politically-motivated actions of several Middle Eastern countries to reduce supply, constituted such a disruption.

The oil price shocks of 1973 and 1974 – often called OPEC I – revealed the short-term inelasticity of demand for oil. This can be explained by the important position petroleum filled in the energy consumption of industrial-

---

[2]Although there was a short-lived recovery in the late 1970s, conditions, particularly in the tanker market, soon deteriorated even further. This chapter focuses on the first stage of the tanker crisis, while the basis for the dismal 1980s will be dealt with in chapter 5.

[3]Norway, Parliament, Stortingsmelding No. 71 (1972-1973), "Langtidsprogrammet 1974-1977" ["Long-term Programme, 1974-1977"], 18.

ised countries and the fact that adaptation to other sources of energy could only occur slowly. Moreover, in many cases the alternatives were either non-existent or extremely expensive. The increase in oil prices dislocated the market, as did uncertainty about the future directions of both price and supply. The shipping industry was hit particularly hard by the price increase. In addition to a sharp reduction in the growth of demand for tankers, the cost of providing transport services increased.

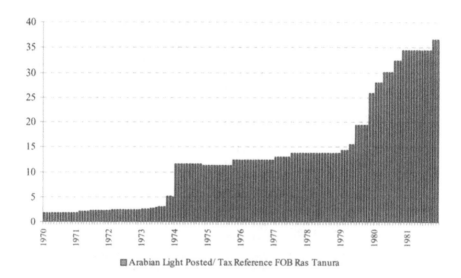

☐ Arabian Light Posted/ Tax Reference FOB Ras Tanura

Figure 3.1: Monthly Oil Prices, 1970-1981 ($ US)

Note: The year on the x-axis refers to January. The figure depicts posted prices, which sometimes deviated from spot prices. For instance, in early 1979, spot prices were as much as US $8 above the posted price. This was one of the reasons for the increases in the posted price later that year; see Joan Edelman Spero, *The Politics of International Economic Relations* (London, 1990), 275.

Source: OPEC, *Statistical Bulletin* (Vienna, 1989), 126-127.

In Fearnley and Eger's *Review 1973*, the consequences of the increased price of oil for the shipping industry were evaluated.[4] Four aspects were emphasised: the demand for tankers decreased initially when the supply of oil was reduced; increased inflation would result from the price increase; the lack of oil was likely to cause problems in bunkering; and increased de-

---

[4]Fearnley and Egers Chartering Co. Ltd., *Review 1973* (Oslo, 1974), 3-4.

mand for relatively inexpensive sources of energy, such as gas and coal, would lead to an increase in demand for tonnage capable of transporting such products. In most of the postwar period the real price of oil had declined because price increases did not match the general rate of inflation. From the beginning of the 1970s, a series of oil price increases were introduced to alleviate this. On 16 October 1973 OPEC announced a two-thirds increase in posted prices. This was supplemented by a further increase, from approximately US \$5 to more than \$11 per barrel, effective on 1 January 1974.

Table 3.1
Major Oil Price Changes, 1971-1980 (\$ US/barrel)

Date	Price before Increase	Price after Increase	Change (percent)
15 February 1971	\$1.800	\$2.180	21.11
16 October 1973	\$3.011	\$5.119	66.96
1 January 1974	\$5.036	\$11.651	131.35
1 October 1975	\$11.251	\$12.376	10.00
1 June 1979	\$15.461	\$19.355	25.19
1 November 1979	\$19.355	\$25.806	33.33
1980 – several changes	\$27.957	\$34.409	23.08

Source: See figure 3.1.

The 1973 and 1974 price increases led to a fundamental structural change in the energy market as oil was transformed almost overnight from a relatively inexpensive and abundant factor of production to a costly commodity. But the full scope of the change was not evident in the short term. According to Oliver Blanchard and Stanley Fischer,

> [i]t was frequently argued that the oil price shock was transitory, likely to last about six months; in fact, it was simply not clear at the time whether the shock was permanent or transitory. Uncertainty about the permanence of the shock must have slowed real adjustment to it, for instance, the adaptation of the capital stock to the higher price of energy.[5]

While these unexpected increases caused immediate problems, the high degree of uncertainty over the long-term effects for shipping was also important in assessing the magnitude of the problems later in the decade. The severity of the shipping crisis was not just a result of an unanticipated reduction in demand but also of a failure to adapt to altered conditions in the freight market. There were three main reasons for this inability to adapt. First, the

---

[5]Oliver Jean Blanchard and Stanley Fischer, *Lectures on Macroeconomics* (Cambridge, MA, 1989), 525.

magnitude of the shock made it extremely difficult to assess the short- and long-term consequences. Second, the more drastic measures that could have been introduced to mitigate the situation, such as mass cancellations of contracts for new tonnage, were not implemented due to the prevailing notion that the shock was temporary. Third, the rigidities in the market for ships, particularly the time lag between contracting and delivery, made adaptation difficult.

In the early 1970s oil consumption was expected to increase by 6.5 percent annually.[6] But the net effect of the OPEC price increases was to reduce consumption in the industrialised countries and to lower expected growth elsewhere. Between 1973 and 1982, oil consumption in the member nations of the Organisation for Economic Co-operation and Development (OECD) fell from 37.4 to 32.2 million barrels per day, and their share of world oil demand dropped from approximately seventy to fifty-seven percent.[7]

The US provides a good example of the discrepancies between actual and anticipated oil consumption. In testimony before the Senate Foreign Relations Committee in late May 1973 William E. Simon, the Deputy Secretary of the Treasury, estimated that domestic demand for oil in 1980 would be about twenty-five million barrels per day. In addition, he predicted that the US would have to increase its share of imported oil from thirty-three to fifty percent of total consumption over the same period.[8] Yet the actual consumption of oil in the US in 1980 amounted to 16.5 million barrels per day, and the share of imported oil increased to more than forty percent. Thus, the US average daily import of oil in 1980 was not the 12.5 million barrels a day expected by Simon but 6.7 million. By 1983, the figure had fallen to less than five million.[9]

Figure 3.2 shows the annual development of oil consumption and exports. While these figures are only partly related to the demand for transport, the large discrepancies between anticipation and reality are striking. Anticipated demand outstripped the actual figures for both oil consumption and exports. The growth of seaborne oil transport was drastically reduced during the 1970s, particularly compared with the period before the price increase.

---

[6]Zenon S. Zannetos, "Market and Cost Structure in Shipping," in Richard L. Gordon, Henry D. Jacoby and Martin B. Zimmerman (eds.), *Energy, Markets and Regulation: Essays in Honor of M.A. Adelman* (Cambridge, MA, 1987), 244-245.

[7]Trevor M.A. Farrell, "The World Oil Market, 1973-1983, and the Future of Oil Prices," *OPEC Review*, IX, No. 4 (1985), 391.

[8]Mohammed E. Ahrari, *OPEC – The Falling Giant* (Lexington, KY, 1986), 37.

[9]Calculations based on figures from British Petroleum, *BP Statistical Review of World Energy 1984* (London, 1984), 7-8 and 16.

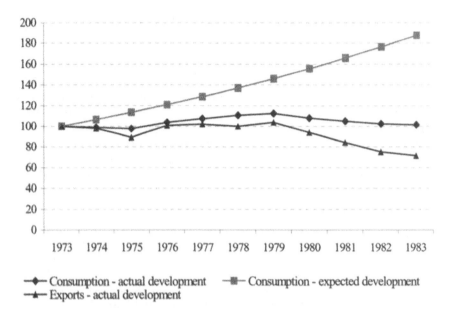

Figure 3.2: Development of Oil Consumption and Oil Exports, 1973-1983 (1973=100)

Source: British Petroleum, *Statistical Review of World Energy 1984* (London, 1984) 8 and 16; and Zenon S. Zannetos, "Market and Cost Structure in Shipping," in Richard L. Gordon, Henry D. Jacoby and Martin B. Zimmerman (eds.), *Energy, Markets and Regulation: Essays in Honor of M.A. Adelman* (Cambridge, MA, 1987), 244-245.

Figure 3.3 illustrates the development of tanker transport demand in the 1960s and 1970s. The strong break from the previous trend is evident when actual developments after 1973 are compared with average growth in the period 1967-1973. The drastic changes would have been even more conspicuous if demand for the transport of oil products had been omitted. Between 1967 and 1973, the average annual increase in the demand for crude oil transport was 19.7 percent, compared with an average reduction of almost one percent in the period 1974-1980. Even though the fall in growth rates was substantial, some sort of reduction should have been anticipated. The closure of the Suez Canal had been one of the reasons for the growth in the period 1967-1973 – as long as the canal remained closed, this was clearly a one-off event.

Perhaps surprisingly, there was no significant decline in the seaborne transport of petroleum and petroleum products in the first year after the dramatic oil price increase. Measured in ton-miles, seaborne transport in 1974 was higher than in 1973, in itself an all-time-high.[10] Despite growing demand, the increase in the transport of crude oil and oil products was considerably less than the historical trend or the expected increase.

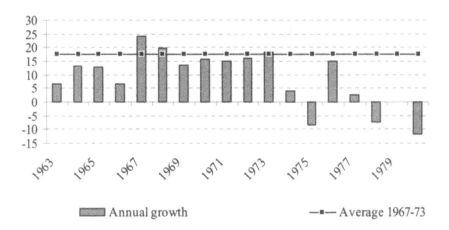

Figure 3.3:   Annual Growth of Seaborne Oil Transport, 1963-1980 (percent)

Note: Calculations in ton-miles; includes transport of crude oil and oil products.

Source: Fearnley and Egers Chartering Co. Ltd., *Review* (Oslo, 1972 and 1981), table 2, transport of crude oil and oil products.

In 1975, there was an absolute reduction in the seaborne transport of crude oil and oil products in a period when the agents – at least before the oil price increases – had expected a considerable increase. It was the first decline since World War II, so demand expectations founded on historical experience would under no circumstances have predicted such a development. Thus, it was in the market for tanker transport that the influence of the oil price increase on the shipping sector first emerged. Due to the embargo by Middle Eastern countries, the expected growth was replaced by a reduction in the demand for tanker transport, instantaneously resulting in depressed freight rates and soon leading to high lay up-rates as well.

The demand for oil is relatively inelastic in the short run. In the longer term it is possible to introduce energy-saving measures and to increase the use of energy sources which have become less expensive relative to oil. A long-term effect of OPEC's cartel pricing policy was therefore an increased

---

[10]There was a small reduction in the seaborne transport of oil and oil products measured in tons, from 1640 million tons in 1973 to 1625 million tons in 1974, but this decrease was more than offset by increasing distances.

reliance upon alternative energy sources. In 1973 oil had accounted for more than fifty-three percent of primary energy consumption in Western Europe, North America and Japan, but by 1982 this had fallen to forty-five percent, with coal, hydro power and especially nuclear energy gaining in importance.[11]

Table 3.2
Countries Registering Significant Increases in Oil Production, 1973-1983

	1973 Production	1983 Production	Increase (Percent)
Mexico	28,000	149,000	428.4
Brazil	8284	16,500	99.2
Egypt	8367	38,000	1804.8
Norway	1575	30,000	1804.8
Denmark	135	2400	1677.8
UK	89	114,500	128,551.7
India	7199	24,000	233.4
Malaysia	4293	18,050	320.5
China	50,000	105,000	110.0
USSR	427,250	618,000	44.6

Note: '000 metric tons.

Source: Trevor M.A. Farrell, "The World Oil Market 1973-1983, and the Future of Oil Prices," *OPEC Review*, IX, No. 4 (1985), 398.

In addition, there was a relative reduction in the consumption of Middle Eastern oil due to increased exploration and the exploitation of wells situated relatively close to major consumption centres. Some of these would have been uneconomical to exploit before the oil price increase (see table 3.2).[12] The annual growth of oil production in Western Europe was 24.1 percent from 1973 to 1982. Over the same period, OPEC's share of world production declined from more than fifty-four percent to approximately a third.[13]

The shift of oil production from OPEC to non-OPEC producers amplified the effects of the falling growth in consumption on transport demand because it led to a reduction in average shipping distances (see figure 3.4). From 1977 onwards, the average shipping distance of oil was reduced as sev-

---

[11]Farrell, "World Oil Market," 394.

[12]The close relationship between the oil price increase and the exploitation of Norwegian oil resources is presented in Norway, Parliament, Stortingsmelding No. 25 (1973-1974), "Petroleumsvirksomhetens plass i det norske samfunn" ["The Role of the Petroleum Business in Norwegian Society"], 6.

[13]Nordine Ait-Laoussine and Francisco R. Parra, "The Development of Oil Supplies during the Energy Crises of the 1970s and Some Questions for the Future," *OPEC Review*, IX (1985), 33.

eral non-OPEC countries increased exploration and exploitation. A larger share of world demand was thus covered by sources relatively close to where the oil and oil products were consumed.

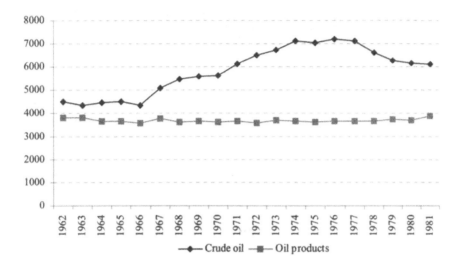

Figure 3.4:   Average Shipping Distance of Crude Oil and Oil Products 1962-1981 (miles)

Source: Fearnley and Eger, *Review*, various years, tables 1 and 2.

The closure of the Suez Canal in 1967 can explain much of the increase in the average shipping distance of oil in the late 1960s. The reopening of the canal in June 1975 did not, however, lead to the expected reduction in distances, particularly for large tankers. The main reason was that the depressed freight rates for large tankers made the use of the canal less advantageous. It was expected that the reopening of Suez would lead to a five percent reduction in dry bulk demand and a maximum ten percent reduction in tanker demand. The actual reduction in demand corresponded to approximately one percent for dry bulk vessels, whereas the reduction in oil shipping demand was negligible.[14] According to Fearnley and Eger's *Review 1975*, only one Very Large Crude Carrier (VLCC) went through the canal in 1975.

The reopening of Suez demonstrates the structural change in the shipping market. Before the canal was closed, approximately three-quarters of cargoes were transported by tankers, but in 1975 liquid cargoes only accounted for about fifteen percent because owners thought it unprofitable to pay the ca-

---

[14]Organisation for Economic Co-operation and Development (OECD), *Maritime Transport 1975* (Paris, 1976), 25.

nal fees to shorten the ballast voyage to the Gulf by ten days when the market was characterised by overcapacity and difficulties finding profitable employment. The most important effect of the oil price increase for the shipping market was the reduced growth, and later absolute reduction, in the demand for tanker transport. But the oil price increase also influenced the cost of providing transport services due to a major increase in the price of bunkers.

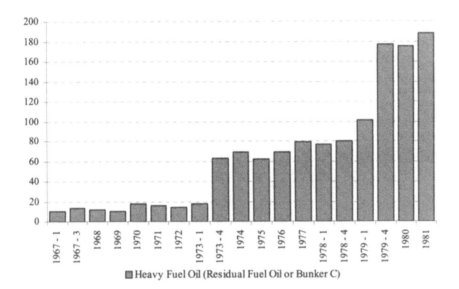

Figure 3.5:   Development of Bunker Prices, 1967-1981 (US $ per ton)

Note: Prices in the bulk market centred upon Northwest Europe, often termed the Rotterdam or ARA (Amsterdam-Rotterdam-Antwerp) market.

Source: Gilbert Jenkins, Martin Stopford and Cliff Tyler, *The Clarkson Oil Tanker Databook* (London, 1993), 269.

Bunkers are available in a variety of quantities and qualities at various locations and may therefore be influenced by local supply and demand conditions. Figure 3.5 thus only gives an indication of bunker prices and conceals large monthly variations. Calculated on a monthly average, the price of "Bunker C" free-on-board at Rotterdam increased from less than US $20 per ton in October 1973 to more than US $120 per ton in December. Thereafter, bunker prices stabilised at around US $70 per ton in 1974.[15]

---

[15]H.P. Drewry Shipping Consultants Ltd., *The Trading Outlook for Very Large Tankers* (London, 1975), 17.

The optimal speed of vessels varies with freight rates and fuel prices. The price of bunkers is one factor determining the shape of the supply curve in the shipping services market. In a market with low freight rates and high bunker prices, shipowners choose to operate their ships well below their maximum speed. Thus, high bunker prices led to an increase in slow steaming to try to improve economic results. Moreover, in the spring of 1974, some vessels had to operate at slow speeds due to the limited availability of bunkers.[16]

The bunker issue is important in determining the cost of providing transport. A group of British shipping consultants tracked a fleet of fifty VLCCs in 1973 and 1974, and their data show the changes after the oil price increase. From 1973 to 1974, annual average speed dropped by eight percent. From an average speed of 14.3 knots in 1973, the decline stabilised at around 12.5 knots in the latter part of 1974, a reduction of about 12.5 percent. Figure 3.6 illustrates the substantial increase in the price of bunkers and its effect. The decreasing difference between bunker prices and total bunker costs – evident from the second quarter of 1974 – shows the savings from slow steaming.

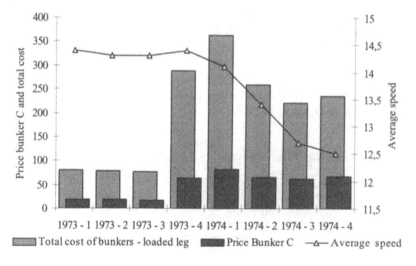

Figure 3.6: Speed and Bunker Costs, 1973-1974 (quarterly)

Note: Persian Gulf-Western Europe only. The bars refer to the left axis and show the price of bunkers in US $ and total bunker costs on the loaded leg in US $ '000. The line refers to average speed, represented on the right axis.

Source: H.P. Drewry Shipping Consultants Ltd., *The Trading Outlook for Very Large Tankers* (London, 1975), 16.

---

[16]*Nordisk Skibsrederforenings Medlemsblad*, No. 497 (1974), 4387-4389.

The quadrupling of the price of bunkers greatly increased the cost of providing transport services. But changes in the price of fuel had different consequences for various operators. Vessels that used a great deal of fuel were adversely affected compared with ships that could be run more economically despite the increase in the price of bunkers. Diesel engines, for example, were more expensive to maintain and install than steam turbine engines but used fuel more efficiently. Accordingly, the cost of operating steam propulsion plants increased considerably more than the cost of running diesels. Another result of the oil price increase was that the use of transport which functioned independently of bunkers, such as pipelines, became relatively more profitable.

The fact that oil companies both supplied and demanded transport capacity makes them difficult to place in any analysis of the causes of the crisis, especially because there was no clear pattern among them. Nevertheless, the oil companies definitely affected conditions in the market. Their policies were important for the growth of the tanker fleet, and by changing their chartering strategies after the market had collapsed they contributed to the spread of the crisis. The close relationship between freight rates and contracting is important in explaining the tonnage surplus. Freight rates were also influential in understanding the timing of charter contracts.[17] In periods when freight rates were high, shipowners were interested in entering into medium- or long-term contracts. Correspondingly, oil companies were willing to commit themselves to charters to reduce the negative effects of any future tonnage shortage. This mechanism can explain the high level of activity in the chartering market in the latter part of 1970 and in 1973.[18]

Table 3.3
Number of "Dirty Fixtures" by Length of Charter, 1970-1977

Year	0-12	12-24	24-36	36-48	48-60	60+	Sum
1970	61	93	206	6	3	21	390
1971	57	8	45	1	25	10	146
1972	42	10	31	5	13	57	158
1973	85	66	170	11	66	9	407
1974	85	14	36	1	9	2	147
1975	49	6	3	0	10	1	69
1976	138	21	10	1	25	7	202
1977	94	10	17	2	9	7	139

Source: D. Glen, M. Owen and R. van der Meer, "Spot and Time Charter Rates for Tankers 1970-77," *Journal of Transport Economics and Policy*, XV, No. 1 (1981), 47; and Fearnley and Eger, *Review*, various years.

[17]A.J. Taylor, "Chartering Strategies for Shipping Companies," *Omega*, X, No. 1 (1982), 29.

[18]Fearnley and Egers Chartering Co. Ltd., *Review 1970* (Oslo, 1971), 17-18.

The large variations in the number of charter agreements are evident in table 3.3. The high level of chartering in periods with high rates affected the level of contracting through two separate mechanisms. First, charter contracts are of direct importance for contracting activity, as these are often used as collateral to finance new buildings. It is thus easier to acquire financing when rates are high, and correspondingly difficult to arrange when rates are low. But the generally soft conditions in the ship finance industry before the crisis reduced the importance of this mechanism. Second, the level of contracting is influenced by owners who act as though rates in the charter market reflect future demand and supply conditions. As a result of the strong correlation between spot and period rates, contracting will be determined by temporary fluctuations rather than the level necessary to create a balance between long-term demand and supply of tanker services.

Figure 3.7 shows the close relationship between freight rates and chartering by oil companies and illustrates the spectacular change in chartering strategy following the oil price increase. In 1973 charter contracts amounting to more than 990 "charter years" were signed compared with a total of 847 "charter years" in the four years 1974-1977. Moreover, the average length of contracts was reduced sharply, from more than twenty-nine months in 1973 to 16.7 months in 1975 and 19.2 months in 1977. In the latter year, more than eighty percent of charters were for one year or less, while in 1973 such short charters comprised fewer than one-quarter of all charters.

Oil companies did not initially change conditions in the market since they did not discriminate against independent shipowners. But they did contribute to a transformation by reserving cargoes for their own ships and those to which they were tied through charters. Initially, the effects of the freight market breakdown affected various shipowners differently. Companies that had their tonnage under long-term contract when freight rates were high at the beginning of the 1970s secured their revenues in the first period after the freight market breakdown.[19] But profitability varied according to when the contract was signed and what kind of contract it was. A voyage charter, in which the shipowner was responsible for bunkers, was far less profitable than a bareboat or period time charter in which the charterer had to cover them.[20]

---

[19]The most profitable charter contracts were medium-length and were signed in the spring of 1970. The highest rates were found in short-term charters, most of which expired in 1972. Ships that had been tied to very long charters in the middle or end of the 1960s were not particularly profitable due to shifts in exchange rates.

[20]Per Gram, "Kansellering av certepartier" ["Cancellation of Charter Parties"], *Norwegian Shipping News*, Nos. 12/13 (1975), 9, claims that one shipowner was willing to *pay* an oil company for a contract from the Persian Gulf to Europe. The rationale was that the oil company under this contract was liable for the bunkering cost, and the shipowner wanted to get the ship to Europe where it would be laid-up.

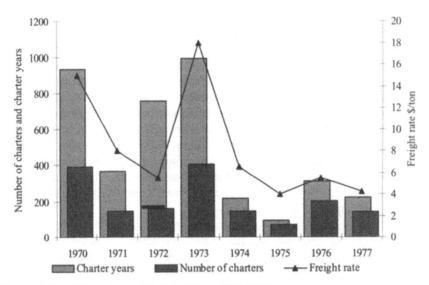

Figure 3.7: Chartering and Freight Rates, 1970-1977

Note: Based on spot rates from the Middle East to Northwest Europe or the US East Coast in US$ per ton; this is represented on the right axis. The number of charters is taken from table 3.3, and the "year" bars represent the number of charters of various length, multiplied by the average duration of the group. Thus, in 1970 the 206 charters of twenty-four to thirty-six months give a total of [(30*206)/12] 515 years. Charters in the "60+" bracket have been assigned a duration of 120 months. The large number of charters signed in 1972 was due to the Sanko deals, which will be discussed later. If the more than fifty contracts of ten years' duration are deducted, the fit between freight rates and "years" in 1972 improves considerably, and the correlation between the rate and the number of charter years increases from an already impressive 0.84 to an astonishing 0.98. The "non-Sanko" level is indicated by the black line in the 1972 bar.

Source: Jenkins, Stopford and Tyler, *Clarkson Oil Tanker Databook*, 250.

After the breakdown of the freight market, it became increasingly common for oil companies not to enter into new long-term charters when the old ones expired. Instead, they covered a consistently larger share of their tonnage requirements in the spot market. The spot market share of oil transport contracts thus rose from ten percent in 1973 to twenty-five percent in 1978.[21]

---

[21]Tormod Rafgård, "Oil Transportation in Tankers – Getting Cheaper and Cheaper," *Norwegian Shipping News*, No. 21 (1979), 29. The figures are largely supported by other sources, although the 1973 figure seems relatively small.

In addition, the number of charter contracts was reduced as a result of an increase in tonnage owned by the oil companies. The share of the tanker fleet owned by these companies increased from thirty-three percent in 1972 to approximately forty percent in 1977; the Seven Sisters alone owned twenty-one percent of the international tanker fleet.[22] Due to increased spot-market chartering and the growth of the fleets owned by oil companies, the share of vessels on charter contracts fell from approximately fifty-five percent before the oil price increase to thirty-five percent in 1978.

One factor that can explain the reduced activity in the charter market was that some of the oil companies had to lay-up their own as well as chartered ships. It is evident that after 1974 the oil companies were hesitant to enter into contracts of more than one year.[23] Another reason that both shipowners and oil companies were more cautious about long-term contracts may have been the increased uncertainty over exchange rates and costs. A Norwegian White Paper claimed that the extensive use of short-term charters in the Norwegian fleet was a result of changes in the policy of charterers, Norwegian cost developments and exchange rate turmoil.[24]

According to Bernhard Abrahamsson, shifts in the use of long-term contracts resulted from structural changes in the international oil market and that it was primarily the oil companies which were reluctant to enter into new contracts as the old ones expired.[25] In addition, an increasing share of transport capacity was bought by small and state-controlled oil companies.[26] Along with increased fragmentation on the demand side, some of the most important oil-

---

[22]H.P. Drewry Shipping Consultants Ltd., *The Tanker Fleets of the International Oil Companies* (London, 1979), 74. The figures vary between different sources; cf., for example, the Drewry figures with those in Mike Ratcliffe, *Liquid Gold Ships. A History of the Tanker 1859-1984* (London, 1985), 161.

[23]D. Glen, M. Owen and R. van der Meer, "Spot and Time Charter Rates for Tankers 1970-77," *Journal of Transport Economics and Policy*, XV, No. 1 (1981). Fearnley and Eger, *Review 1975* (Oslo, 1976), 12, claims that only two of the large American oil companies undertook any significant period chartering in 1975.

[24]Norway, Parliament, Stortingsmelding No. 23 (1975-1976), "Om sjøfolkenes forhold og skipsfartens plass i samfunnet" ["On the Conditions for Seamen and Shipping's Position in Society"], 57-58.

[25]Bernhard J. Abrahamsson, "International Shipping: Developments, Prospects and Policy Issues," *Ocean Yearbook 8* (Chicago, 1987), 172. As the lack of chartering was connected to the price of long-term charters, one might just as well claim that owners were "reluctant" at the prevailing rates.

[26]Bernhard J. Abrahamsson, "Merchant Shipping in Transition," *Ocean Yearbook 4* (Chicago, 1983), 176.

producing nations reduced the length of their exploitation contracts with the oil companies. The share of internationally-traded oil based on long-term contracts fell from ninety percent in 1973 to forty-two percent in 1979.[27]

The reduced use of long-term contracts and the increasingly important position of smaller oil companies may also explain the decline in charter agreements. It might be expected that charter activity would increase when small companies, owning little tanker tonnage, needed to secure their delivery of oil, but this does not appear to have been the case. Whereas the seven majors on average covered approximately seven percent of their tonnage requirements in the spot market in 1978, the leading non-majors covered 24.5 percent of their needs there.[28] But this difference in chartering strategy was not necessarily a result of structural changes in the market. A plausible explanation may be that the relatively small oil companies did not perceive a lack of tonnage as likely in a market characterised by considerable tonnage surplus and accordingly chose to buy relatively cheap transport capacity in the spot market.

The behaviour of the oil companies, characterised by a strong reduction of activity in the charter market, was one factor that contributed to the spread of the crisis to owners who initially had escaped the market collapse. After the oil price increase, shipowners in the period market were unable to renew or negotiate new contracts when their current ones expired. They were therefore forced to lay-up vessels or accept the abysmal rates in the spot market. The high degree of concentration on the demand side, where a small number of companies bought a large share of the transport capacity, made it easy for the oil companies to keep track of their competitors' chartering policies.

Due to the massive tonnage surplus, the need for oil companies to insure against tonnage shortages through long-term charters was reduced. This explains why an increasing share of transport was undertaken at extremely low rates in the spot market. The fact that relatively few oil companies demanded most of the tonnage supplied by independent shipowners may be important in explaining the spread of the crisis. The Norwegian shipowner Erling Dekke Næss had previously claimed that the oil companies chose to enter into charter contracts individually, thereby forcing rates down.[29] But he does not present any evidence to substantiate this, and there is no support for it in the literature.

[27]Olav Bjerkholt, Erik Offerdal and Steinar Strøm (eds.), *Olje og gass i norsk økonomi* [*Oil and Gas in the Norwegian Economy*] (Oslo, 1985), 56.

[28]Calculations are unweighted averages based on the percentage breakdown of the different types of oil company-owned, period and spot tonnage in Drewry, *Tanker Fleets*, 73.

[29]Erling Dekke Næss, "Tankfartens problemer og utsikter" ["The Problems and Prospects of Tanker Shipping"], Kristofer Lehmkuhl Lecture, Norwegian School of Economics and Business Administration, 1965.

The change in chartering strategy spread the crisis to owners who had traditionally operated in a relatively "prudent" manner in the charter market. Moreover, the chartering policies may have lengthened the crisis as well if before the price increase the oil companies consciously or unconsciously gave independent shipowners an economic motive to undertake large investments. While there is little doubt that the high rates in late 1972 and 1973 stimulated independent shipowners to invest in new vessels, it is impossible to prove that this was some sort of plan to lead to a tonnage surplus. On the other hand, the increase in the size of the oil companies' tanker fleets during the 1970s shows that they did not hesitate to take advantage of low second-hand prices, which more than compensated for their relatively low contracting activity in 1972 and 1973 when their share of new buildings was smaller than their share of the world tanker fleet.

Another way in which the oil companies contributed to market imbalance was by failing to cancel new building orders when overcapacity became evident. Some companies which had drastically overestimated their future transport requirements did cancel such orders, but as the OCED remarked, "it is hard to see oil companies being sufficiently altruistic as to pay cancellation fees for the privilege of chartering tonnage to fill the requirement."[30]

The development of the demand for tanker tonnage was unfavourable. Both the price increase and the policies of the oil companies affected tanker owners negatively. Yet these points cannot fully explain the extent of the calamity. Another reason for the severity of the crisis was supply-side expansion. The effects of negative demand were amplified through the actions of owners, shipbuilders, banks and governments. Their choices – both before and after the oil price increase – made a difficult situation even worse.

The maritime economist Martin Stopford has summarised the causes of the international shipping crisis of the 1970s. "Looking back to 1973, the influence [of these institutions] is clear – the banks lent too much, the shipbuilders built far more shipyards than long-term demand forecasts justified and shippers issued time charters which underpinned the false sense of optimism."[31] Accordingly, a prominent topic in any analysis of the crisis must be the reasons for the strong growth of the supply side.

The following section focuses on three groups: shipbuilders, shipowners and financial institutions. All played crucial roles – shipyards built the ton-

---

[30]OECD, *Maritime Transport 1975*, 106.

[31]Martin Stopford, "Challenges and Pitfalls of Maritime Forecasting in a Corporate Environment," in Siri Pettersen Strandenes, Arnljot Strømme Svendsen and Tor Wergeland (eds.), *Shipping Strategies and Bulk Shipping in the 1990s* (Bergen, 1989), 42. When presented in this manner, the quote seems to omit the actions of the most important group of actors: the shipowners. Stopford's topic, however, was the factors that made owners order new tonnage.

nage; shipowners ordered it; and financial institutions facilitated the investments. The effects of politics and government intervention will not be dealt
with separately but will be discussed in relation to the influence of the various
agents.

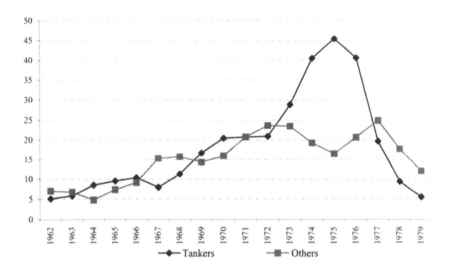

Figure 3.8:   Deliveries of New Buildings, 1962-1979 (million dwt)

Source: Fearnley and Eger, *Review*, various issues, table 4.

The strong growth of the world fleet was closely connected to the
massive increase in global shipbuilding capacity after World War II. From
1945 to 1975 there was extremely strong growth in the actual and potential
supply of new buildings (see figure 3.8). Along with the increased capacity,
the large demand for tonnage caused the world's order books to increase from
eighteen million gross registered tons (grt) in 1960 to more than 133 million
grt in 1974.[32] The amount of new tonnage launched increased from almost seventeen million deadweight tons (dwt) in 1965 to 61.8 million dwt in 1975.

The increased capacity of shipbuilding in the postwar period was the
result of an increase in efficiency as well as growth in several countries. The
increased efficiency was based upon improvements in technological processes
and better organisation of production. One of the most important features was
an increase in the maximum size of ships, which represented an opportunity
for the expansion of shipyards and technological change in the industry.

---

[32]Anthony Slaven, "Management Policy and the Eclipse of British Shipbuilding," in Fred M. Walker (ed.), *European Shipbuilding: One Hundred Years of Change*
(London, 1983), 82.

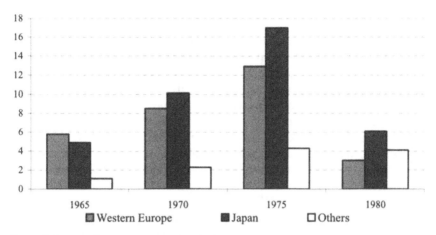

Figure 3.9: Production in World Shipbuilding, 1965-1980 (million grt)

Source: Lloyd's Register of Shipping, reprinted in Cees de Voogd, "Public Intervention and the Decline of Shipbuilding in the Netherlands," in Randi Erstevåg, David J. Starkey and Anne Tove Austbø (eds.), *Maritime Industries and Public Intervention* (Stavanger, 1995), 242.

Parallel with the capacity increase, there was a spatial shift in the world's shipbuilding industry (see figure 3.9). The Japanese share increased from twenty percent in 1960 to approximately fifty percent eight years later, mainly at the expense of Great Britain and other European countries. The Japanese expansion was largely a result of the country's policy of industrial integration in which crucial sectors, such as shipbuilding and vehicle production, were nurtured for their employment potential, their important backward linkages to heavy industry and the availability of a large international market. But the increased market share of Japanese shipbuilding was to some extent "paid for" by subsidisation and generous financing terms.

The shipbuilding industry's role as a key labour-intensive industry was reflected in its central position in domestic industrial, employment and regional policies. Shipbuilding was responsible for a large share of manufacturing employment in many countries, such as Norway and Sweden.[33] Moreover, it was an important employer in certain regions and cities. In Sweden in the mid-1970s, shipyards were the major employer in some relatively small towns, such as Landskrona and Uddevalla, where shipbuilding employment accounted for thirty-nine and fifty-two percent of manufacturing jobs, respectively. But it was also important in larger cities, accounting for twenty-five

---

[33]Helge W. Nordvik, "The Norwegian Shipbuilding Industry – The Transition from Wood to Steel 1880-1980," in *ibid.*, 201.

percent of manufacturing employment in Gothenburg and nineteen percent in Malmö.[34]

The strong international competition in shipbuilding led to the extensive use of subsidies and cheap financing to attract new building contracts. In several countries the authorities used mechanisms such as export credits, direct subsidies and investment support. The result of the high level of national subsidisation was a vicious circle in which the various shipbuilding nations constantly increased their support in order to strengthen their competitive position in the international market.[35] The outcome can be illustrated by a statement which ironically was used to justify national support to the Swedish shipbuilding industry. According to an OECD report, "[t]he profitability problems of the Swedish shipbuilding industry are to a great extent due to the State subsidies given in other countries."[36]

The variety of measures through which the shipbuilding industry was subsidised makes it virtually impossible to quantify their effects. In fact, researchers are unable even to agree on whether subsidies in certain countries were high or low in international perspective. In 1969 the OECD formulated an "Understanding on Export Credits" to try to secure a higher degree of real competition in shipbuilding and to neutralise the effects of the various support measures. Under this policy, the participating countries were obliged to comply with maximum limits for direct or indirect support; financing could not be on terms better than a six-percent interest rate, with an eight-year down payment and twenty percent in cash.[37] At the time, the OECD countries built more than ninety percent of new tonnage.

---

[34]Lars Berggren, "The Effects of the Shipyard Crisis in Malmö, Southern Sweden," in Randi Erstevåg, David J. Starkey and Anne Tove Austbø (eds.), *Maritime Industries and Public Intervention* (Stavanger, 1995), 194-204.

[35]For a short introduction to the use of subsidies in shipbuilding and shipping in the unregulated period in the first decades after the Second World War, see S.G. Sturmey, *British Shipping and World Competition* (London, 1962), 188-209. An instructive introduction to the mechanisms influencing the shipbuilding industry in the period after Sturmey's focus is de Voogd, "Public Intervention," where the emphasis is on the Dutch shipbuilding industry.

[36]OECD, *The Industrial Policies of 14 Member Countries* (Paris, 1971), 318. Subsidisation is rational in Sweden, if it is regarded separately. But collectively this behaviour is irrational. The sum of the measures was an increase in supply, which reduced profitability and which, when the agents were acting rationally on an individual basis, formed the basis for further support.

[37]Norway, Parliament, Stortingsproposisjon No. 101 (1976-1977), "Om tiltak på skipsbyggingssektoren" ["On Measures for the Shipbuilding Sector"], 19-20. The minimum interest rate was raised to 7.5 percent in 1971.

The support given to shipbuilders in all the major building nations contributed to the over-ordering and consequently exacerbated the international shipping crisis. The favourable conditions, mainly the provision of cheap financing, led to a shift in demand from second-hand tonnage to new buildings. The subsidised financing contributed to an increase in contracting by reducing the demands on the ships' revenue. Because the purpose of cheap financing was to create an artificial competitive advantage for the country's shipbuilders, it was tied directly to the contracting of new tonnage in the nation providing this subsidy. Due to this intervention in the market, shipping capital was not allocated to investments that would have provided the highest socio-economic profit. Instead, subsidies functioned as an incentive to invest in a market where there already were signs of overcapacity.

As a result of the large number of subsidies available, shipowners who wished to acquire new tonnage did not have to pay the true market price. This led to tonnage growth that exceeded the level necessary to maintain a long-term equilibrium in the sector. Shipyards did not determine their capacity on the basis of the amount of tonnage that the shipping sector could absorb but instead built what they could induce shipowners and oil companies to order. Subsidies played an important role in this regard, and their size correlated negatively with the prevailing market conditions, since it was more important to use them to generate activity in periods when shipbuilding was slack. The shipyards therefore gave shipowners incentives to order more tonnage when market prospects dictated a reduction in new building.

The role of shipbuilding in an analysis of the shipping crisis depends upon the period examined. Before 1973, shipbuilding in part created its own demand by providing relatively favourable financing. Yet the effect of subsidies immediately before the crisis should not be overemphasised. In a boom period the authorities had relatively little incentive to subsidise shipowners, as the favourable conditions in the shipping market were sufficient to secure new orders. Nevertheless, due to the generous terms that evolved during the scramble for market share in the 1960s, there was a certain degree of subsidisation even when the market was good. The terms laid down in the OECD Understanding represented the "minimum assistance" that shipowners could expect to receive. It did succeed in preventing cut-throat competition, since prior to its adoption it had been possible to get eight-to-ten-year credits at 5.5 percent. Nevertheless, even the OECD terms represented an interest rate subsidy that led to an inflated level of contracting.

Although aid to shipbuilding influenced the high contracting level before 1973, its effects were even more dramatic after the freight market collapsed. The reluctance – particularly by governments – to reduce shipbuilding capacity must be understood in terms of the industry's social significance. Initially, the shipping market was affected through the yards' reluctance to accept the cancellation of new building orders. Moreover, their demand that tanker contracts be converted to other types of tonnage contributed to the spread of the crisis from the tanker sector to other parts of the industry.

A shipowner who is interested in investing in new tonnage will, all things being equal, choose a new building instead of second-hand tonnage if the financing terms are more favourable. This will be particularly relevant in periods with low freight rates. Empty building berths mean that the new vessel may be delivered relatively quickly, as opposed to half a decade during booms, thus reducing the difference between the time of operation for the new building and the second-hand vessel. Furthermore, short order books tend to encourage governments to secure orders through subsidies.

In the longer term, growth was exacerbated because shipbuilders finishing new vessels competed for new contracts. As John B. Yolland claimed, "[i]t is an established fact that the world shipbuilding over-capacity is causing governments of all countries with shipyards to ignore many agreements in order to attract employment and alleviate social and economic problems."[38] Additionally, some non-OECD countries, notably South Korea, Taiwan and Brazil, which were not bound by the agreements, expanded their shipbuilding industries, adding to the competition and the shipyard overcapacity.

In most shipbuilding nations, the authorities took a great interest in maintaining building capacity even after the tonnage glut and the demand reduction had led to a breakdown in the tanker market. According to Lewis Johnman, "Europe's response to the international shipbuilding crisis was to subsidise on the expectation of an improving market and to reduce the capacity on the assumption that it would not."[39] In the 1970s, at least, the former strategy was more prominent than the latter. In Sweden, the shipbuilding industry received SKr twelve billion in state support between 1977 and 1979, in addition to subsidies that reduced the price of steel.[40] This can be compared with the SKr two billion in subsidies given to all sectors of Swedish industry in the seven previous years. Moreover, the Swedish Guarantee Fund was willing to

---

[38]John B. Yolland, "Ship Finance and Euro-markets," *Maritime Policy and Management*, VI, No. 3 (1979), 175.

[39]Lewis Johnman, "Public Intervention and the 'Hollowing Out' of British Shipbuilding: The Swan Hunter Closure," in Erstevåg, Starkey and Austbø (eds.), *Maritime Industries and Public Intervention*, 223. The focus of this paper was the crisis in shipbuilding when the new buildings ordered at the beginning of the 1970s were cancelled or delivered, but several of the situations discussed were parallel to those in the period before the breakdown of the markets for shipping and ships.

[40]The enormous transfer of money was motivated by the fact that shipyards were extremely important for manufacturing employment in a number of Swedish cities. See Bo Stråth, "Industrial Restructuring in the Swedish Shipbuilding Industry," *Labour and Society*, XIV, No. 2 (1989), 106.

lend up to seventy-five percent of the value of vessels for which no order had been received.[41] This building "on speculation" illustrates the lengths to which shipyards and governments were willing to go to avoid a sudden reduction of production and employment. Nevertheless, before the oil price increase and the freight market breakdown, this kind of speculative building was unnecessary. Finding shipowners to invest in new shipping capacity was an easy task.

An analysis of the criteria used by shipowners to make investment decisions can help us understand the massive growth of the world fleet in the period around the shipping crisis. There are a number of relevant factors. First, shipowners tended to conduct only superficial analyses before signing contracts for new vessels. More specifically, they placed an inordinate emphasis on historical and current freight levels. Related to this was that in the long growth period before the crisis, rapid expansion had generally been quite profitable. By merely extrapolating from this trend, most of them could justify new building orders. Third, many seemed to ignore the fact that the lag between contracting and delivery of vessels, which can be particularly lengthy in boom periods when building berths are full, means that the market could (and did) change drastically before the ships entered operations. Fourth, many were seduced by the increasing competition in the ship financing industry, which lowered bank margins and eased credit terms. Moreover, this was exacerbated by the abundance of capital in the Eurodollar market which was related to American economic policy and contributed to high liquidity and low real interest rates. These low rates made contracting seem more attractive due to the low returns on alternative investments and the low cost of carrying debts. Finally, there was the impact of taxes, which almost everywhere made contracting preferable to paying levies to governments in years with high profits.

The years 1970-1973 were characterised by extremely high contracting activity for tankers. Figure 3.10 shows clearly the enormous increase in contracting from 1965 until the peak in 1973. Tanker tonnage contracted in 1973 was twice as high as the previous year, which was itself an all-time high. The share of total contracts accounted for by tankers increased from almost forty-seven percent in 1969 to more than eighty-four percent in 1972.

The strong increase in tonnage ordered illustrates the extraordinary development in the early 1970s. This expansionary strategy is even more conspicuous if we compare the contracting figures with the size of the existing fleet. As figure 3.11 shows, tanker tonnage on order amounted to less than fifty percent of the existing fleet at the beginning of 1970. This was high relative to the mid-1960s but low compared with the greater than ninety percent

---

[41]Jan Kuuse, "The Relations between Swedish Shipbuilding Industry and Other Industries 1900-1980," in Kuuse and Anthony Slaven (eds.), *Development Problems in Historical Perspective* (Glasgow, 1980), 233; and "The Many Aspects of the Shipbuilding Crisis," *Norwegian Shipping News*, No 17D (1976), 6-15.

level at the beginning of 1974. One reason for the increasing size of the order books was the growing size of vessels on order. In 1965 the tankers on order were on average 123 percent larger than the typical vessel in the current fleet. This figure increased to 165 percent by 1974. Moreover, new orders as a share of the current fleet, which had been less than fifteen percent in 1965, peaked at more than fifty-five percent in 1973. One of the main causes was the increase in demand for tanker capacity and the high freight rates, particularly in 1970 and 1973. The fact that contracting persisted at a relatively high level in 1974, when freight rates were plunging, may seem to undermine the high correlation between freight rates and contracting. But much of the contracting in 1974 was by Middle Eastern oil companies which wanted to supplement their increasing control of production with greater control over transport. The orders were thus motivated by political and structural issues rather than economic considerations.[42] Moreover, it is likely that some of the contracts reported in 1974 were signed before the freight market collapsed.

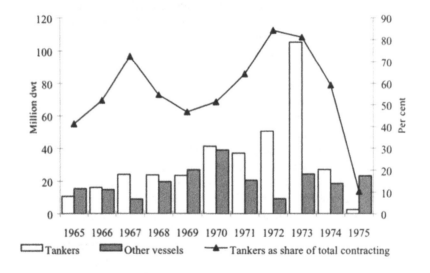

Figure 3.10: World Contracting and Tanker Share, 1965-1975 (million dwt)

Note: Combination carriers included as "other vessels." If these had been counted as tankers, the tanker share of contracting would have peaked at 87.5 percent in 1972 and at about eighty-four percent the following year.

Source: Fearnley and Egers Chartering Co. Ltd., *Review 1976* (Oslo, 1977), 18.

---

[42]For an introduction to the growth of the fleet of the oil exporting countries, see "The Middle East Fleet – A Threat or An Opportunity?" *Norwegian Shipping News*, No. 8B (1975).

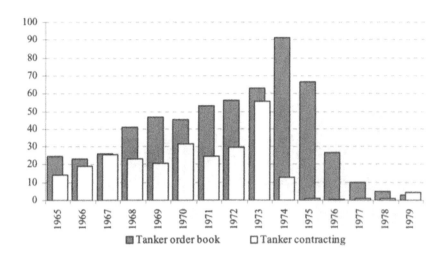

Figure 3.11:   Tanker Contracting and Order Book, Percent of the Tanker Fleet, 1965-1979

Note: The figures for contracting refer to contracting undertaken during the year, whereas figures for the order book refer to the situation on 1 January.

Source: Fearnley and Eger, *Review*, various issues, tables 9 and 10.

Figure 3.12 is based on orders for tankers above 100,000 dwt. It illustrates the relative importance of three groups – oil companies; independent shipowners with employment secured; and independent shipowners ordering tonnage which had not been fixed on charters. The figure does not take into account, however, variations in contracting between different periods. Moreover, the data refer to the number of orders and do not reflect the tonnage ordered by the various groups. Yet in 1970 and 1971 oil companies accounted for approximately forty percent of the new building orders. This was slightly higher than their share of the world fleet, but considerably above their proportion of contracting in the two subsequent years. Their large contracting at the start of the decade may have been motivated by a fear of future capacity shortfalls relative to the anticipated high demand for transportation. Their share of new building contracts in 1972 and 1973 was considerably smaller, and the fall was especially noticeable when the independent shipowners' tanker contracting ballooned in the latter part of 1972. The number of new oil company contracts for tankers above 100,000 dwt more than halved between the first and fourth quarters of 1973, while the number of new building contracts signed by independent shipowners increased by almost three-quarters. Another conspicuous feature was the high share of unfixed new building orders signed between the second quarter of 1972 and the last quarter of 1973. Indeed, this was the period with the highest contracting activity.

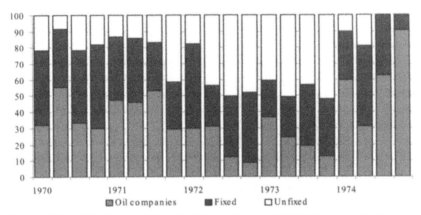

Figure 3.12:   Distribution of New Buildings, Percent, 1970-1974 (quarterly figures)

Note: Quarterly data on new building contracts over 100,000 dwt. The year on the x-axis refers to the first quarter.

Source: Petter Dragesund, "Kontraheringsatferd i tankmarkedet" ["Contracting Behaviour in the Tanker Market"] (unpublished høyere avdeling thesis, Norwegian School of Economics and Business Administration, 1990), 49.

Figure 3.13 illustrates the considerable growth of the world fleet from the beginning of the 1960s until the end of the 1970s. The demand for new tonnage, which is the basis for supply-side growth in the market for shipping services, was closely related to the shipowners' assessment of the profitability of investing in additional capacity. The result of such an evaluation depends to a great extent on the investor's expectations concerning the future costs and revenues associated with the project and his evaluation of the risk involved. Andreas Vergottis uses an asset-pricing approach to describe the demand for new tonnage.[43] In his framework, the price shipowners are willing to pay for a new vessel is equal to the discounted stream of expected profits in connection with the new building, plus a price subsidy, the value of cheap finance and the value of tax advantages.

---

[43]Andreas Rokos Vergottis, "The City University Econometric Model of the Shipping Markets," in Strandenes, Svendsen and Wergeland (eds.), *Shipping Strategies*, 24-38.

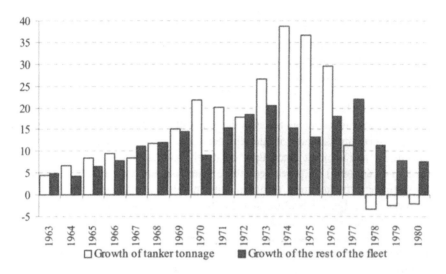

Figure 3.13: Annual Growth of Tankers and the Rest of the Fleet, 1963-1980 (million dwt)

Note: Due to the fluctuating level of scrapping, there is a deviation between fleet growth and new tonnage delivered (figure 3.1).

Source: Fearnley and Eger, *Review*, various years, table 3.

    The discounted stream of expected profits is a measure of future earnings deriving from the investment. The size of this part of the equation is determined by the shipowner's expectations about prospective rate and cost levels. The assessment of potential income (represented by the freight rates) and costs (which encompass operational commitments and interest payments on borrowed capital) embodies a considerable degree of uncertainty. In particular, uncertainty about expected profits is particularly high in volatile markets, such as shipping, in periods of economic, financial and political instability, like the early 1970s.

    One of the cash flows included in the discounted expected profit stems from the eventual sale of the ship, either to new owners or a scrapyard. This may constitute a considerable part of the overall income. A vessel's discounted expected profit is thus influenced by the shipowner's expectations about future supply-and-demand conditions. Moreover, due to the high correlation between second-hand prices and freight rates, the discounted value of the vessel's sale is closely related to the general market conditions when the vessel is sold.

    A price subsidy will reduce the cost of the investment because part of the price of the ship is paid by agents other than the shipowner. As the value of the subsidy grows, the revenue needed for the investment to be profitable is reduced.[44] Due to the subsidisation of shipbuilding, shipowners do not bear the real costs of increasing transport capacity.

The value of cheap finance may or may not be part of the subsidisation and has two important connotations. First, shipowners tend to evaluate fewer alternative investments than do general corporations. Second, cheap financing provided by shipyards for new ships often acts as a kind of indirect subsidy. It has been claimed that shipping appears more sensitive to interest rates than other investments because the main alternatives for shipowners often are short-term investments in the financial market. Long-term non-shipping investments are avoided since their relatively low liquidity reduces financial freedom of action. This perceived lack of alternatives implies that deficiencies in the finance market may result in uneconomical investments in ships.[45] If shipowners receive misleading signals from the financial markets about the alternative value of their capital, the level of shipping investments may be elevated. On the other hand, if the interest rates on borrowed capital are artificially low, shipowners may be encouraged to increase their borrowing because the revenue requirements to service the debt are relatively low.

Cheap financing, such as loans provided by shipyards on preferential terms, may be part of a price subsidy. This will be the case when the loans offered have better terms than those the shipowner would be given by the financial market. The results of this kind of subsidisation will again be a reduction in the revenue necessary for a shipowner to service the debt and a fleet that is larger than it would have been had the shipowner borne the actual costs of building the vessel.

The effect of tax advantages can also influence the expected profit from a shipping investment. In some countries, the tax system may be arranged so that part of the associated risks or costs is borne by government. If this is so, the profit and risk evaluations of the shipping company will be altered, and it is possible that this will lead to investments that would not have occurred without the influence of the tax system. It is evident that the combination of low real interest rates and large depreciation allowances led to extremely low, or even negative, real after-tax interest rates in the early 1970s. This promoted contracting and fleet expansion.

Period rates are strongly correlated with the spot-market rate. Consequently, rather than mirroring long-term market trends, they reflect short-term

---

[44]When demands on the ship's revenue decrease, the shipowner is often willing to pay more for it. The accumulated payment will be the shipowner's share plus the subsidy, but the shipowner will only take his own investment into account when evaluating profitability. As long as the subsidies are tied directly to contracting new buildings, their existence leads to an increase in the amount of tonnage ordered.

[45]This point is emphasised in Victor D. Norman, "Shipping Problems – Has the Market Mechanism Failed?" *Norwegian Shipping News*, No. 7 (1976), 27.

fluctuations in supply and demand.[46] This has important implications for fleet development and can explain why the volume of tonnage ordered reached record-breaking proportions in 1973.

Returning to the asset-price approach, an important determinant of the expected stream of profits is the level of freight rates. If period rates are used to make investment decisions – that is, if shipowners choose to contract due to high rates in the period market – long-term investment decisions will be influenced indirectly by temporary demand fluctuations in the spot market. This mechanism is simple: the shipowner's decision to contract ships is determined by his assessment of the future demand and supply for transport which to a large degree is influenced by the current level of freight rates in the period market. The rates a shipowner might obtain by fixing the new building on, say, a five-year time charter, may be viewed as an "objective" indication of the freight rate level over the period. By ordering a new vessel, and by chartering it on the available terms, the shipowner can fix the expected revenues for a given period. The uncertainty regarding the discounted stream of expected profits is then basically reduced to uncertainty about the shipping and second-hand ship markets when the charter terminates and uncertainty about the movements of exchange and interest rates.[47]

As fluctuations in charter market rates are highly correlated with spot market rates, the current state of the shipping market is an important factor in shipowners' contracting decisions. This is the element that Stopford referred to as "time charters which underpinned the false sense of optimism."[48] Victor Norman has found that variations in the spot rate between 1962 and 1971 "explain" eighty-four percent of the variations in contracting.[49] The highest correlation is between fluctuations in contracting and freight rates for medium-term charters. But due to the close relationship between period and spot market

---

[46]See, for instance, Ib Erik Eriksen and Victor D. Norman, *Ecotank – modell for analyse av tankmarkedenes virkemåte* [*Ecotank – Econometric Model for Tanker Companies*] (Bergen, 1976); and Siri Pettersen Strandenes, *Kontrahering og salg av norske tankskip 1963-76* [*Contracting and Sales of Norwegian Tankers 1963-76*] (Bergen, 1979).

[47]For some types of charters there is the added uncertainty over the trend in costs.

[48]Stopford, "Challenges and Pitfalls," 42.

[49]Victor D. Norman, *Internasjonal Sjøtransport* [*International Sea Transport*] (Bergen, 1973); and Norman, *The Economics of Bulk Shipping* (Bergen, 1979), 18. In "Har det vært for lett å drive internasjonal skipsfart i Norge?" ["Has Shipping Been Too Easy an Enterprise in Norway?], *Internasjonal Politikk*, No. 1B (Oslo, 1979), 182, Norman claims that current revenue can explain seventy percent of the fluctuations in the contracting volume of Norwegian tanker owners in the period 1963-1976.

rates, the influence of the latter on contracting is considerable. Periods with high freight rates tend to coincide with a considerable over-ordering of ships.[50] This was evident immediately before the freight market collapse.

One type of new building contract of particular interest in an analysis of contracting in the period around the shipping crisis is the speculative contracts by independent shipowners who had not yet secured long-term charters for their vessels. In an analysis of the contracting of tankers of over 100,000 dwt between 1970 and 1974, Petter Dragesund found that speculative contracting was considerably more evident in 1972 and 1973 than at other times in this era.[51] He claims that the main causes of this were the short investment horizons of independent shipowners, erroneous expectations and general insecurity about future demand.[52]

Another way to explain the importance of current freight rates is through the concept of adaptive expectations. Applied to shipping, this posits that shipowners place inordinate emphasis on current and historical rate levels. The result is that contracting is relatively high when demand and freight rates are high or increasing. This may occur even if the cause of such conditions is temporary. One example is the demand increase in connection with the 1967 Suez crisis. For shipowners, one of the most important results of the closure of the canal was that it became increasingly profitable to build VLCCs and Ultra Large Crude Carriers (ULCCs). These vessels could only be operated profitably when oil shipments were relatively long distance.[53]

---

[50]Over-ordering implies a level of contracting which is so large that it causes future imbalances between supply and demand.

[51]One particular type of speculative contracting was to order ships because of an expected increase in new building prices. Unless such contracts are based on expectations about an increase in the long-term freight rate level, they tend to destabilise the market; see Petter Dragesund, "Kontraheringsatferd i tankmarkedet" ["Contracting Behaviour in the Tanker Market] (unpublished høyere avdeling thesis, Norwegian School of Economics and Business Administration, 1990).

[52]The short-term investment horizon can explain contracts designed to reap the benefits of rising new building prices. Increased insecurity about demand stemmed from uncertainty about oil prices and international monetary relations.

[53]Customers of the Norwegian shipbrokers Fearnley and Eger were among those who invested in VLCCs and ULCCs at an early stage; see Lewis R. Fischer and Helge W. Nordvik, "Economic Theory, Information and Management in Shipbroking: Fearnley and Eger as a Case Study, 1869-1972," in Simon P. Ville and David M. Williams (eds.), *Management, Finance and Industrial Relations in Maritime Industries: Essays in International Maritime and Business History* (St. John's, 1994), 23.

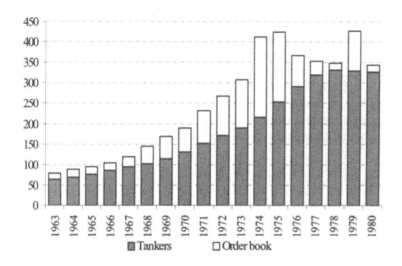

Figure 3.14: Existing Tanker Fleet and Order Book, 1963-1980 (million dwt)

Source: Fearnley and Eger, *Review*, various years, tables 3 and 9.

Figure 3.14 illustrates both the expansion of the fleet of existing and ordered vessels and the growth of the tanker order book after the good years 1967, 1970 and 1973, when the volume of tanker tonnage on order increased by sixty-nine, thirty-six and sixty-five percent, respectively.[54] The basis for this activity was the large fluctuations in spot-market rates caused by short-term shortages of transport.

Historical events, such as the Yom Kippur War, will lead to high spot and charter rates. Moreover, such incidents initially trigger a boom in freight rates and ordering. Due to high rates in the period market, contracting increases despite often unchanged prognoses about future demand and supply. A long-term disequilibrium between supply and demand may occur because the high demand and freight rates are unsustainable in the long run. One reason that this may be true is that the situation that initially led to high freight rates has been altered.

Since independent shipowners have limited resources to monitor the market, they often rely on external sources. Oil companies and large shipowners, on the other hand, undertake more sophisticated analyses. One way that small shipowners can obtain such information is through the freight level in the period market, where high freight rates suggest that the major shippers and shipping companies believe that a future capacity shortage is likely, while low

---

[54]Because these figures represent the order books at the beginning of the year, the effects of the freight market peaks occurred in 1968, 1971 and 1974.

rates may reflect a perceived long-term overcapacity. Independent owners may therefore see high rates in the period market as a signal to order more ships. If this is based upon a long-term charter, it is rational from the shipowner's perspective since the charter secures long-term revenue. While there is a small risk concerning the charterer's ability to fulfil the agreement, in general the cost of an oversupply of tonnage is borne by the charterer. But when high freight rates lead to unsecured contracting, the situation is somewhat different, particularly with regard to the shipowner's risk. Such contracting was one of the main causes of the problems in the Norwegian shipping industry.[55]

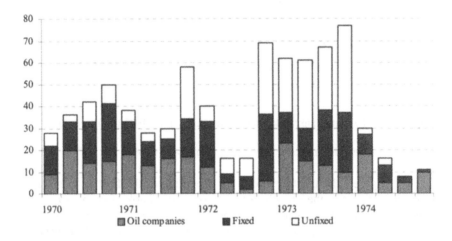

Figure 3.15: Number of New Building Contracts, 1970-1974 (quarterly)

Note: Vessels above 100,000 dwt only. Years refer to the first quarter.

Source: Dragesund, "Kontraheringsatferd i tankmarkedet," 49.

        Figure 3.15 shows the number of contracts signed for tankers above 100,000 dwt between 1970 and 1974. Oil companies bore the risk for eighty percent of the contracts signed in 1970, 1971 and the first half of 1972, either by contracting vessels or through charters. This means that the oil companies had agreed to pay for the employment of the vessels. Accordingly, less than twenty percent of the tonnage contracted during this period was unfixed, although for vessels contracted in the second half of 1972 and in 1973, almost half had not secured employment. As a result, the risk stemming from overcapacity was increasingly borne by the independent shipowners. Indeed, the number of unfixed new building orders increased from less than ten in the first

---

        [55]The relationship between contracting and charter rates which forms the basis for this argument is elaborated upon in Norman, *Economics of Bulk Shipping*, 17-19.

three quarters of 1972 to an average of more than thirty in the following five quarters. Since the independent tonnage contracted in 1972 and 1973 was less likely to be based on long-term employment, a large share of the risk was transferred from the oil companies back to the shipowners.

Another notable feature is that the boom in independent contracting took place from the fourth quarter of 1972 onwards – spot market rates had trebled between April and September. The effects of this on the contracting by independent shipowners evidently did not take long to materialise. Indeed, the contracting in late 1972 is typical of the "herd behaviour" often observed in shipping. As the Norwegian shipbroker R.S. Platou remarked in its 1972 annual review,

> [i]n October, the contracting market exploded into activity...In trying too understand the activity that took place in October and November...we believe it is necessary to appreciate that the general pessimistic outlook for the future...had been turned to cautious optimism...The constant flow of reports on the American energy crisis indicated the need for a future vast increase of foreign oil supply.[56]

The close relationship between freight rates and contracting, leading to periodic over-ordering, was doubtless important in explaining the considerable tonnage increase in the early and mid-1970s. The over-ordering contributed to an increase in the breadth, length and depth of the shipping crisis.

The breadth of the shipping crisis, i.e., the number of those affected, can be directly attributed to the large supply of new vessels entering the market. Shipowners who had ordered large tankers without secure employment in advance, or who were operating their existing fleet in the spot market, were hit particularly hard. The fact that their vessels had to be laid-up or offered in the market at low rates was true even for shipowners who had refrained from contracting during the boom. The pool of inactive tonnage meant that any increase in the demand for transport would attract laid-up vessels rather than contribute to more than a marginal rate increase.

As oil companies turned increasingly to spot chartering and short-term contracts after 1974, even those shipowners who had secured their ships' employment through medium- or short-term charters were hit by the crisis. The average duration of the charters entered into prior to the collapse was approximately thirty months, with a majority running for between two and three years. As these expired, it became increasingly difficult to enter into profitable new contracts. Independent shipowners were forced to lay-up their vessels,

---

[56]R.S. Platou, *The Platou Report 1972* (Oslo, 1973), 17-18.

accept the miserable rates in the spot market or enter into charters at levels only slightly above operating costs, leaving little room to recoup capital costs.

The length of the crisis was also influenced by the high level of contracting. The life of a ship traditionally has been estimated at approximately twenty years. This implies that a large share of the ships ordered during the boom at the start of the 1970s were still available for service throughout most of the 1980s and 1990s.[57] Most of the tanker fleet was between thirteen and nineteen years old in 1992, reflecting the high volume of contracting in the early 1970s and the large deliveries from the middle of that decade.

The over-ordering also influenced the depth of the crisis, particularly the amount of tonnage laid-up. It is unlikely that freight rates would have reached a higher level with a smaller degree of surplus tonnage, since in theory the marginal freight rate will be attained regardless of whether the surplus is ten or two hundred ships.[58] New tonnage was introduced into a market where there was no hope for profitable employment, and it was not uncommon for new buildings to go straight into lay-up. The share of the world fleet laid-up was high enough to have a profound effect on most participants in the tanker market. Even the oil companies, which could secure cargoes for their own fleets, had to lay-up some vessels. This was a result of overestimating future transport needs, particularly with their large new building orders in 1970 and 1971.

The growing imbalance between tanker demand and supply is crucial to an understanding of the international shipping crisis of the 1970s. The contracting boom at the beginning of the 1970s is important to explain the strong growth of the supply side, and the effects of the periodically high freight rates can explain much of why they were ordered. But one significant question remains: why did the tanker fleet continue to grow even after the surplus had become evident? This can be explained by the lag between contracting and delivery and the fact that cancellations were costly. It might be expected that shipowners, who adapt relatively swiftly to increases in freight rates by increasing their contracting, would choose to cancel contracts when freight rates plunged. Yet a considerable period of time passed before the owners of trans-

---

[57]The long-term influence of the contracting boom is evident from surveys of the age composition of the tanker fleet; see Terje Lensberg and Heine Rasmussen, *A Stochastic Intertemporal Model of the Tanker Market* (Bergen, 1992), 9, for a survey of the situation in 1990.

[58]It may be argued, however, that the capacity potential of the inactive fleet is important; if ten vessels are laid-up, the demand increase necessary for an upward movement in freight rates is far smaller than if two hundred are inactive. The high level of contracting is nevertheless important for the depth of the crisis, since supply growth led to rates that were so low that even the most efficient vessels were laid-up.

port capacity and new building contracts were fully able to fathom the extent of the crisis.

The lag between contracting and delivery varied with the size of the order book and consequently with the state of the shipping market (see figure 3.16). In periods with high freight rates and contracting, it might take as long as five or six years from the time vessels were contracted until they were delivered. Some large tankers ordered shortly before the freight market collapsed at the end of 1973 were thus scheduled to be delivered as late as 1978-1979.[59]

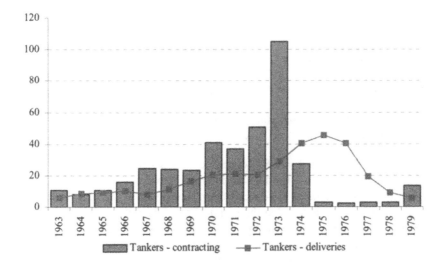

Figure 3.16:   Contracting and Delivery of Tankers, 1963-1979 (million dwt)

Source: Fearnley and Eger, *Review*, various years, tables 4 and 10.

The amount of tanker tonnage delivered reached record-breaking levels in both 1974 and 1975, when market conditions suggested that both short- and long-term capacity reductions were necessary. The fleet continued to grow despite the bleak prospects. More than forty million dwt were delivered in 1976, even though a report from a British shipping consultant more than a year previously had stated that "there will be continuing over-supply of VLCC-tonnage throughout the remainder of the 1970s."[60] In 1976 the OECD esti-

---

[59]Fearnley and Eger, *Review 1973*, 5.

[60]Drewry, *Trading Outlook for Very Large Tankers*, 50.

mated that a balance could be restored by 1983.[61] Some new building contracts
were cancelled, but the number was far too small to achieve a balance between
supply and demand (see table 3.4). In 1977 Fearnley and Eger claimed that the
"yard industry has...faced a total cancellation volume of 64 million dwt, of
which 15 million dwt were registered as conversions."[62] The exact volume of
cancellations is difficult to estimate due to the existence of new building options which may or may not have been declared, the treatment of conversions
and the existence of new building orders that were not "genuine." The figure
usually cited is in the region of sixty million dwt.[63]

Table 3.4
Estimates of Cancellations, 1974-1976

	Tankers	Other Tonnage	Conversion
1974	7-9 million dwt		
1975	26 million dwt		9-12.5 million dwt
1976	9.5-13 million dwt	1.2 million dwt	

Source: OECD, *Maritime Transport*, various years: and Fearnley and Eger, *Review*,
various years.

Although unfixed vessels contracted by independent shipowners comprised slightly more than a third of all contracts between 1971 and 1974, they
accounted for about half of the cancellations (see figure 3.17). But these were
insufficient to restore balance to the sector. There are three main reasons that
the cancellations remained relatively low despite the changed conditions.

First, it took considerable time before shipowners realised the extent
of the crisis. Although the transformation was both sudden and dramatic, representing a fundamental break with developments in previous decades, some
time passed before it was recognised that this was a permanent shock rather
than a transitory glitch. In 1976, the OECD remarked that:

> [e]ven after the impact of the oil crisis there was a consider-
> able time-lag before it was realised that a long-term depres-
> sion on the tanker market would probably take place, and
> only recently have shipowners, oil companies, shipbuilders,

[61]OECD, *Maritime Transport 1975*, 93.

[62]Fearnley and Egers Chartering Co. Ltd., *Review 1976* (Oslo, 1977), 5.

[63]See Ludwig Beth, Arnulf Hader and Robert Kappel, *25 Years of World
Shipping* (London, 1984), 36; and OECD, *Maritime Transport 1976* (Paris, 1977), 80.

bankers, and under force of circumstances, governments also started to look seriously at solutions to adjust the situation.[64]

It is evident that even though expectations regarding future market conditions had changed, the break was so fundamental and the projections were associated with such a large degree of uncertainty that the agents were unwilling to reduce the capacity to counter a "worst case" scenario.

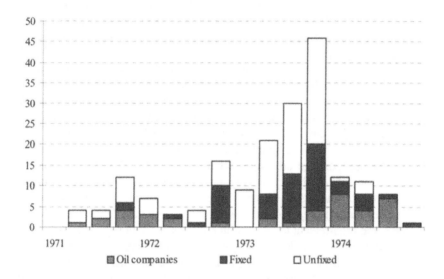

Figure 3.17: Cancellations of New Buildings by Time of Contracting, 1971-1974

Note: Quarterly data for contracts of more than 100,000 dwt. Years refer to first quarter. Figures include new building contracts which were later cancelled with no information about when the cancellation occurred.

Source: Dragesund, "Kontraheringsatferd i tankmarkedet," 60.

Second, cancellation did not appear to be a viable alternative for some agents with large new building orders on their hands. Shipowners who had secured employment for their new buildings through long-term charters were reluctant to pay cancellation fees since they were assured income, at least temporarily. If they expected the market to improve in the long term, cancellation implied that they would have to forego future profit opportunities. Similarly, it

---

[64]OECD, *Maritime Transport 1975*, 100.

would usually be unprofitable for the oil companies to pay cancellation fees if they later had to buy the necessary transport capacity on the open market.[65]

Third, the cancellation of new building contracts imposed severe penalties on shipowners. Those looking for outright cancellation would usually have to pay a penalty of at least ten to twenty percent of the contract price, but sometimes as much as fifty percent.[66] Some vessels in the later stages of the production process were impossible to cancel. As shipyards realised that new orders would be increasingly hard to secure, cancellations became even more difficult to arrange. Conversion of tonnage was an alternative, but it implied a penalty as well. Although the fees were smaller for conversions, the spread of the crisis and the subsequent deterioration of conditions outside the tanker sector made this a less attractive alternative.

The sometimes exorbitant level of cancellation fees can be illustrated by the case of one Norwegian shipowning company which had ordered four 150,000-dwt tankers at an average price of $32.5 million for delivery in 1976-1977. When it approached the shipyard to cancel one or more of the contracts, "the demand from the yard was approximately 2/3 of the building price."[67]

The expansion of shipyard capacity and the eagerness of shipowners to acquire new tonnage are important in explaining the increasing supply of transport capacity. A third group that contributed to the oversupply of tonnage was the banks and other financial institutions. Although a subsidised yard could offer interest rates at least as attractive as those available in unsubsidised markets, the requirement to pay within eight years could represent such a drain on resources and cash flows that government financing had to be combined with long-term debts. Moreover, the fact that government financing in some nations, notably Japan and Germany, became more closely linked to the domestic currency added an element of exchange rate risk to the equation.

Since the governments of the major shipbuilding nations and institutions with little experience in ship financing funded a large share of these investments, the demands for collateral and equity may have been lower than they might otherwise have been. The banks, which had a long experience with shipping and therefore a competitive advantage in evaluating these projects, were initially seen as conservative relative to the state-supported institutions and those banks that entered the shipping sector with large amounts of capital

---

[65]As figure 3.10 shows, though, some oil companies, such as those which had seriously overestimated their transport requirements, found it profitable to cancel new building contracts.

[66]In some instances the contract was cancelled by mutual agreement.

[67]Archives of the Norwegian Shipowners Association, folder 6/B/K/75, Krisen 1975, IV, 011075-301175, letter from a Norwegian shipowner, 24 November 1975.

after 1967. The specialised shipping banks often regarded long time charters as crucial for long-term profitability and security, and they were thus accused of forcing shipowners to enter into unprofitable time charters. But as competition intensified, even the established financial institutions lost some of their traditional prudence.

The basis for the growing importance of financial institutions in the shipping market was the increased size, and hence higher price, of vessels. Due to the increase in capital intensity, the financing of vessel acquisitions from a company's internal resources became increasingly difficult. Furthermore, relatively low real interest rates, and the fact that increasing debt capital reduced the amount of the shipowners' own resources at risk, made borrowing a favourable alternative to equity financing.

Peter Stokes claims that "the volume of finance available has not been a constraint...If anything, the reverse was the case, and shipowners' inherent lack of self-discipline has been dangerously encouraged by the profusion of funds proffered by the banks and credit agencies."[68] One reason for this large supply of capital was the influx of new banks into ship financing in the 1960s, particularly after the Suez crisis boom in 1967. The strong growth of international trade and the volume of seaborne transport in the first postwar decades created a need for fleet expansion, with a corresponding demand for the financing of shipping investments. In addition, the profits made by the banks and financial institutions that facilitated these investments were sufficient to attract the interest of institutions without specialist shipping departments. The result was an increase in the shipping portfolios of traditional shipping finance institutions such as Hambros Bank, Chemical Bank and Chase Manhattan, as well as the entry of new agents with little experience into the shipping sector.

The increased competition in ship financing led to a reduction of the margins for financial institutions. In addition, they also competed on the terms they offered. One result was that the traditional conservative terms were abandoned; it became increasingly easy to finance a larger share of a new building with loans. Peter Stokes calls the period 1967-1973 "the ship lending boom" and claims that the market "had lost all sense of restraint" as "some deals were transacted in which the borrower actually received finance for more than the cost of the vessel."[69] The extent of such transactions should not be overstated, but it is clear that the financial industry was quite liquid and that the terms offered to shipowners became increasingly generous.

The ship financing industry changed during the 1960s. First, London became increasingly important at the expense of New York. Second, American

---

[68]Peter Stokes, *Ship Finance: Credit Expansion and the Boom-Bust Cycle* (London, 1992), 115.

[69]*Ibid.*, 24.

economic and foreign policies created a booming market for dollar financing abroad. In 1963 the American authorities introduced the Interest Equalization Tax on foreign borrowing, which raised the costs of foreign bonds sold in the US market. The result was the rapid growth of a Eurodollar market, where American banks were among the most eager participants (see figure 3.18). This market's liquidity was secured by the expansionist policies of the US government, particularly after 1969, which made an ample supply of dollars available and greatly increased world liquidity.

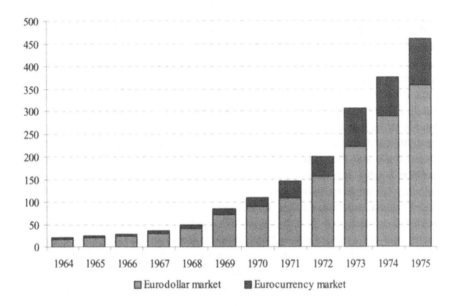

Figure 3.18:   Growth of the Eurodollar Market, 1964-1975 (US $ billion)

Note: The growth of the Eurodollar market after 1975 was enormous, mainly due to the influx of Petrodollars. By 1980 the market had more than trebled relative to 1975, and in 1985 it amounted to almost $3000 billion, of which American dollars accounted for three-quarters.

Source: Spero, *Politics of International Economic Relations*, 42.

Access to undemanding financing was the result of a variety of factors. At the international level, the US balance-of-payments deficit contributed to the abundance of capital in the Eurodollar market. The massive dollar outflow resulted in the spread of inflationary tendencies from the US because European countries had to accept the dollars that flooded the international financial markets in the late 1960s and early 1970s in order to maintain the fixed exchange rates of the Bretton Woods system. Moreover, government subsidies to shipbuilding and the entry of new and relatively inexperienced agents into the ship financing market also increased the capital available to shipowners. Competition reduced bank margins, led to lower interest rates and generated more generous conditions for borrowers. This benefited shipowners looking to outside sources to finance fleet renewal or expansion. The abundance of capital was accompanied by increasing rates of inflation, which made real interest rates relatively low (see figure 3.19).

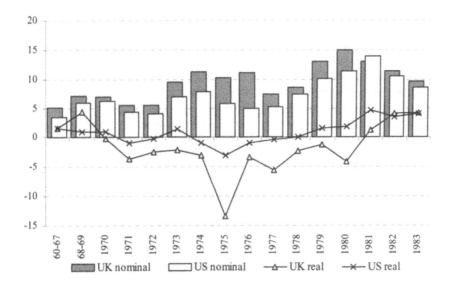

Figure 3.19: Nominal and Real Short-term Interest Rates, 1960-1983

Note: Monthly averages based on American certificates of deposit and British three-month interbank loans after 1975 and three-month treasury bills prior to 1975. To calculate real rates, nominal rates have been divided by the ratio of the GDP implicit price index for the current to the previous year. Figures for 1960-1967 and 1968-1969 are annual averages.

Source: OECD, *OECD Economic Outlook - Historical Statistics 1960-1983* (Paris, 1985), tables 10.7 and 10.8.

   The pattern of real interest rates contributes to an explanation of the large investments in shipping in 1972 and 1973. What occurred in shipping was closely correlated with the general business cycle, implying that freight rates were often high during inflationary periods. This is logical, since high inflation leads to low real interest rates because nominal rates adjust relatively slowly to increasing inflation. The real interest rate reflects an assessment of current relative to future values and affects the saving and investment decisions of shipowners. High inflation reduces the attractiveness of the short-term money market, where nominal interest rates have not increased sufficiently to account for higher inflation. Moreover, low real interest rates increased the attractiveness of investments in shipping where expected yields were relatively higher.

For shipowners with surplus capital, a short-term investment in the money market is a common alternative to investing in shipping. Long-term investments are seldom regarded as an alternative because they tie up resources for a considerable time, thus reducing the room to manoeuvre. Low real interest rates make it less profitable for shipowners to look to the short-term money market and increase the profitability of investing in new transport capacity.

The state of the international capital markets and the ship financing industry in the early 1970s is thus important in understanding the high contracting volume and the magnitude of the shipping crisis. The large amount of shipping capital available was not only a reflection of the abundance of capital in international markets but also a result of the good returns from shipping investments prior to the crisis. In this respect, financial institutions succumbed to the same temptations as the shipowners to whom they granted loans: they too placed an inordinate emphasis on the strong historical development of the shipping sector and failed to anticipate any changes in market conditions.

Just as soaring freight rates attracted shipping investors, the increased competition and abundance of capital overcame some of the traditional prudence of financiers. Expectations of continuing transport growth and booming markets led banks and other financial institutions to abandon traditional requirements, such as long-term charters, as prerequisites for granting loans for vessels. While this development was gradual, it increased from the late 1960s onwards. Established shipowners with relatively safe portfolios initially were granted mortgages for more speculative investments. Later, gearing was increased for all kinds of companies, and the conservative collateral requirements were waived. Peter Douglas claims that the cause of the problem was that "the concept of standard credit terms for creditworthy buyers of ships degenerated, in the boom atmosphere of the time, into the idea of standard credit terms for almost anyone who wanted to buy ships."[70] According to one industry insider, "we tended to make up the rules as we went along...[F]ew lending officers actually had any experience of problem loans in shipping."[71]

The development of ship prices is important to understand the softening of financing conditions. In the late 1960s and early 1970s, vessels were seen as "floating real estate" because favourable market conditions and infla-

---

[70]Peter S. Douglas, "Financial Policies for the Shipping Industry," in Inger Rydén and Christopher von Schirach-Szmigiel (eds.), *Shipping and Ships for the 1990s* (Stockholm, 1979), 169.

[71]Quoted in Stokes, *Ship Finance*, 25.

tion normally enabled shipowners to sell their assets for more than the original price, often after a considerable period of profitable operation. Figure 3.20 shows the pattern of second-hand vessel prices in the late 1960s and early 1970s and illustrates the basis for optimism in the financing sector. Why should banks hesitate to finance investments which retained their value after several years, sometimes even after the effects of inflation? An 80,000-dwt tanker built in 1966/1967 could be sold for more than the original price in every year between 1969 and 1973; if it were purchased in 1968, it could be sold at a profit of almost 150 percent two years later.[72]

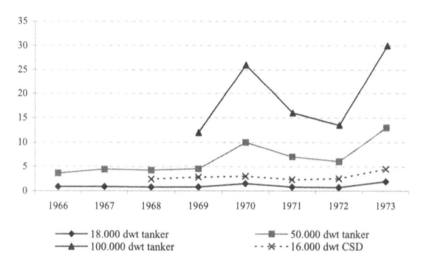

Figure 3.20: Second-hand Prices, Tankers and Dry Cargo Ships, 1966-1973 (US $ million)

Source: Fearnley and Egers Chartering Co. Ltd., *Review 1972* (Oslo, 1973), 16.

Shipping was a favourable sector for Eurodollar mortgages due to the high capital requirements and the excellent market for second-hand vessels which made investments relatively liquid. Nevertheless, due to high tanker prices, loans often required syndicates since the capital needed was too large even for the major banks. Sometimes there were as many as twenty participants in a financing scheme organised by a lead bank, usually one of the traditional shipping banks. On the one hand, syndication reduced risk exposure; but on the other, it increased the possibilities for inexperienced financial institutions to gain a share of the supposedly lucrative ship financing business.

---

[72]Fearnley and Egers Chartering Co. Ltd., *Review 1972* (Oslo, 1973), 16, also contains additional information on the types of vessels and estimates of values.

The amount of money lent to the shipping sector increased substantially in the late 1960s and early 1970s. Concern about the state of the shipping industry resulted in the establishment of the International Maritime Industries Forum in 1976, which in turn established a study group to quantify the sector's debts. Its estimates in the middle of 1976 revealed mortgage debts of about US $25 billion on the existing world fleet of tankers, combination carriers, medium and large bulk vessels and oil rigs. In addition, there were debts amounting to more than $12 billion related to vessels and rigs on order (see table 3.5). More than half the debt could be attributed to tankers of more than 200,000 dwt. Otto Norland, a London-based Norwegian banker and the chairman of the study group, estimated that US $4.3 billion of the loans were related to tankers without fixed employment and that this figure could increase to US $10 billion by 1980 if all repayments of principal were deferred due to the delivery of new unfixed vessels and the termination of existing charters.[73]

Table 3.5
Mortgage Debts in Shipping, July 1976 (Million US $)

Year of Delivery	Tankers 200+	Tankers 80-200	Tankers <80	Comb. Carriers	Bulk 60+	Grand Total
1969	120	12	4	5	3	144
1970	309	40	50	54	28	481
1971	530	63	74	172	113	952
1972	1050	108	120	484	217	1979
1973	1802	212	338	642	333	3327
1974	2725	712	370	440	335	4582
1975	4122	1681	829	424	561	7617
1976 (Q1&2)	2113	965	562	274	420	4325
Total	12,771	3784	2347	2495	2010	23,407
On order	4381	1593	1864	869	2550	11,257

Note: These figures are estimates and may deviate from the actual debt by ten percent for individual figures and five percent for aggregates.

Source: Peter Stokes, *Ship Finance: Credit Expansion and the Boom-Bust Cycle* (London, 1992), 54.

The booming Eurodollar market and the financial sector's positive expectations facilitated the eagerness of shipowners to expand their fleets. Neither party was thinking irrationally given the historical development of shipping, the path of ship prices and the prevailing market prospects. But the demand side played havoc with these expectations, and it became obvious that the desire of financial institutions to increase their presence in the ship finance

---

[73]Stokes, *Ship Finance*, 55.

market had led to a laxity which became costly once the market shifted from growth to stagnation.

The preceding analysis has presented some of the aspects that facilitated the large investments before the freight market collapse. One additional element which should be considered is the effect of the tax system in the major shipping nations. While the direct links between taxation policy and large shipping investments are examined more thoroughly in the chapter on Norwegian shipowners, the importance of the tax system for the high international level of contracting warrants a short introduction here. The favourable tax treatment of shipping occurred in part because traditional maritime nations were experiencing increased competition from low-cost flags of convenience. By offering a generous tax regime, governments could alleviate some of this competitive disadvantage.

In several of the most important shipping nations the design of the tax system made large investments attractive for shipowners with high revenues, thus contributing to the close relationship between rates and contracting. The combination of low real interest rates and advantageous tax regimes often resulted in a negative post-tax real interest rate on shipping investments. The result was that some contracting may have been undertaken for tax reasons; shipowners found it favourable to order tonnage which would not have been contracted in the absence of incentives built into the tax system.

Through the taxation of income, investments or capital, governments may contribute to a change in the strategic behaviour of shipowners. This will be the case if the design of the tax system induces the person or company taxed to adapt their business decisions to the tax regime. In the case of investments, the situation is analogous to the influence of the authorities on the shipbuilding sector, where subsidies and soft financing reduced the revenue necessary for investments to be profitable.

If the tax system reduces the actual price shipowners have to pay to acquire tonnage, the volume contracted will increase. The most blatant example of this kind of influence is the direct investment grants of twenty to twenty-five percent available to British shipowners.[74] In international perspective, the tax treatment of depreciation was a particularly important source of distortion. Some of the costs of new investments could be offset by tax reductions due to liberal depreciation rules, making it profitable to have high investments in years with high revenues. In the UK the investment grants were combined with "free depreciation," which enabled British shipowners to write off the cost of a new ship in the first year of operation if sufficient profits were earned.[75]

---

[74]Richard O. Goss, "Rochdale Remembered," *Maritime Policy and Management*, XXV, No. 3 (1998), 220-221.

[75]Ronald Hope, *A New History of British Shipping* (London, 1990), 462. According to Goss, "Rochdale Remembered," 220, the Chamber of Shipping, in connec-

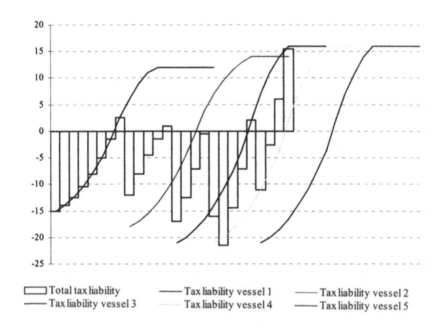

Figure 3.21:   Shipping Operation and the Tax Effect of Reinvestment

Note: While tax liability in the latter part of the period has been deleted to
                improve visual presentation, it will be considerable unless new invest-
                ments are made. The figure ignores the effect of revenues from the sale
                of vessels.

Source: Courtesy of the author.

        The effect of the tax system on contracting can be illustrated by the
hypothetical situation of a new shipowning company. For a typical shipping
investment, the development of costs and revenues follows an S-shaped curve.
Depreciation and high interest payments lead to negative results in the initial
period, which generally spans the first years of the investment. Later, net
revenues from operations become positive, and the company becomes liable
for taxes. This liability may, however, be reduced or eliminated by new investments which lead to interest payments and depreciation which can be deducted from revenue. Figure 3.21 illustrates the possibilities to reduce total tax
liabilities by continuous investment in new tonnage. Only when the company
refrains from investing do taxes at the company level become positive. The
columns in figure 3.21 show that total tax liability is negative for some time

---

tion with the 1970 Committee of Inquiry into British Shipping, sought to enlarge the
depreciation allowance to include money already covered by the investment grant.

after the company invests in a new vessel but turns positive when income exceeds depreciation and other deductions. By investing in new vessels, total tax liability may be reduced due to depreciation. If the vessels earn a considerable profit, it is necessary to expand the fleet increasingly rapidly.

One important element, which is only partly captured in the figure, is that tonnage acquisitions tended to become more expensive with the increase in ship size and vessel specialisation. This amplified the positive effect of investments on taxes. Moreover, the figure does not take into account the fact that the revenue, and thus tax liability, of the existing fleet varies with fluctuations in the market. Accordingly, the desire to reduce the taxable income of the existing fleet through investment in new tonnage is particularly great when the freight market is good.

The fact that tax-induced contracting is particularly attractive in periods with high revenues makes the distorting effects of the tax system particularly harmful. Over-contracting takes place principally when high rates induce shipowners to expand their fleets. The influence of the tax system could of course be the sole reason for this, if all contracting was motivated by the desire to reduce taxation on high incomes in boom periods. A more likely explanation, however, is that contracting is based on high expectations regarding future freight rates, and that the large contracting in boom periods is exacerbated by the effects of the tax system.

The massive increase in tanker transport capacity is one of the keys to an understanding of the international shipping crisis. Several elements contributed to the high level of contracting prior to the freight market collapse. The shipbuilding industry had increased its capacity considerably and was able to supply a large amount of new tonnage, albeit with a lag between contracting and delivery. Shipowners were willing to order vessels. An expansive strategy had proved profitable in the growth of the first postwar decades, and it was hard to envisage a sudden change in the development. The financial community competed for market shares in the ship financing market, and due to developments in the international economy, a large volume of capital was available for investment. A combination of all these factors is important in explaining the contracting boom. The amount of tonnage contracted and the size of the order book reached unprecedented levels, spurred by the high level of period and spot-market rates, particularly from the middle of 1972 onwards, which resulted in a favourable view of future tanker demand. Actual demand, however, did not meet expectations and represented a clear break with historical development.

The tanker crisis was triggered by the oil price increases of October 1973 and January 1974 and the subsequent cartel policy followed by OPEC. The magnitude of the crisis can to a large extent be explained by the actions of participants in the tanker market in the period around the freight market collapse and the strong, unexpected reduction in the growth of demand. In particular, the relationship between shipyards, shipowners and bankers, and the expectations each had, were crucial to subsequent developments.

Two aspects of the strategic behaviour of the shipyards, shipowners and bankers were particularly important. First, prior to the freight market breakdown their strategies reflected an anticipation of sustained strong growth in demand for tanker transport. Due to the oil price increase, this growth failed to materialise. Second, the inertia after the market collapse can be explained by the fact that everyone involved acted as though the problems were transitory. The effect of the strategic decisions, both prior to and after the oil price increase, was that the volume of tanker tonnage grew considerably despite stagnating demand. Indeed, the shipping crisis was the result of considerable discrepancies between expected and actual developments and reflected the fact that a transition to the altered situation in the tanker market was difficult to implement.

The tanker market was where the difficulties first became evident, and in 1974 and 1975 the problems were largely referred to as "the tanker crisis." But the interrelationships between various segments of the shipping market meant that the problems soon spread more widely.

Chapter 4
The Contagion and Short-term Effects of the Crisis

In the postwar period the various segments of the shipping market were increasingly served by specialised vessels. But the reduction in the number of general-purpose ships does not imply that the various shipping segments were isolated from each other. Indeed, the problems in the tanker sector spilled over relatively rapidly into other parts of the shipping market.

Two mechanisms on the supply side – the inflow of combination carriers and the eagerness of shipyards to build new bulk ships – explain the spread of the crisis from tankers to the dry bulk sector. At the same time, it is likely that a recession would have occurred in the latter even without the tanker problem due to a slowdown in world manufacturing. The situation was analogous to the tanker market: the recession would have been particularly strong because those involved expected continuing demand growth.

In the dry bulk market, the immediate effect of the oil price hike was a strong increase in the price of bunkers. Again, this particularly affected owners operating relatively fuel-inefficient vessels, such as older ships or large ones with turbine engines. Fortunately, due to differences in the size distribution, turbine-driven vessels were less important in the dry bulk than in the tanker sector.

A more fundamental effect of the oil price increase was the long-term influence on the demand for dry bulk transport. The increased price of oil amplified the downward business cycles that were already evident in the world economy. Due to the strong correlation between the demand for dry bulk transport and manufacturing, the recession had negative consequences for bulk transport demand.

The oil price hike was one of the main causes of the international economic recession of the 1970s. It is possible, however, to find some signs of disequilibrium in the world economy before oil prices were raised, such as problems in international monetary relations. The world economy was struck by high inflation, primarily in connection with increasing prices for raw materials and wages. In addition, the stability and predictability of international trade relations were reduced because of a loss of confidence in some of the world's major currencies.

Due to disequilibrium in the international economy, a reduction in the strong economic growth that characterised the twenty-five years after the end of World War II could have been expected. The Organisation for Economic Co-operation and Development (OECD) even suggested that "1974 was likely

to be a recession year even without the oil crisis."[1] The oil price hike rein-
forced both inflation and recession in the industrialised countries. As governments sought to counter the former, recessionary forces became even more
prominent. Table 4.1 shows the reduction of economic and employment
growth and the increase in inflation.

Table 4.1
OECD: Main Economic Indicators, Annual Averages (Percent)

	1965-1973	1973-1979	1979-1985
Real Output Growth	4.9	2.8	2.1
Inflation	5.1	9	7.7
Employment Growth	1.1	1.1	0.6

Source: OECD, *Structural Adjustment and Economic Performance* (Paris, 1987), 54.

The manipulation of oil prices by the Organisation of the Petroleum
Exporting Countries (OPEC) aggravated a negative trend in the international
economy and led to considerable changes in the terms of trade. Oil-exporting
nations experienced a dramatic increase in foreign earnings, whereas countries
that were net importers of oil developed large trade deficits. The result was a
transfer of purchasing power and monetary reserves. The oil-producing nations
spent some of their new-found wealth on consumption, particularly of industrial goods from Germany, the US and Japan. Moreover, a considerable share
of the Petrodollars was returned to the industrialised countries as deposits in
the Eurodollar market.

A stable world economy would have reacted to the oil price increase
in a different manner. An examination of the business cycle revealed that the
most important industrial nations were heading towards recession. The oil
price increase amplified the effects of this, as in practice it constituted a considerable "tax" on the consumption of oil products in the industrialised world.[2]
Moreover, economic policies were tightened to control the inflation stimulated
by the boom of the early 1970s and the subsequent increase in the price of one
of the most important factors of production.

In January 1974 the finance ministers of the leading industrialised
nations agreed to avoid policies that would exacerbate the problems posed by

---

[1]Organisation for Economic Co-operation and Development (OECD), *Towards Full Employment and Price Stability* (Paris, 1977), 70.

[2]Herman van der Wee, *Prosperity and Upheaval. The World Economy 1945-
1980* (London, 1986), 494, argues that "[t]he raising of oil prices had important consequences for the world economy and the working of the international monetary system...
It was inevitable that total demand in the industrialised countries would fall, and this
largely explains the seriousness of the world recession of 1974-1976."

the oil price increase. Yet as the shipping analysts H.P. Drewry pointed out six months later:

> since that meeting, Italy has resorted to import restrictions, the UK has had a deflationary budget, the US has pushed up interests to record heights, Denmark has imposed large purchase tax increases to stem imports, Japan has launched a vigorous export campaign...and the new French government is preparing an austerity programme [i.e., deflation] matched by a drive to accelerate exports.[3]

These responses aggravated, rather than alleviated, a difficult situation.

After 1973 there was substantial growth in the supply of tonnage in the dry bulk sector, in the short as well as the long run. The cause of supply-side growth initially was the introduction of combination carriers previously employed in the tanker market. This redeployment was a direct effect of the tanker market collapse. In the longer run, there was a large increase in the supply of dry bulk tonnage due to the conversion of tanker new building contracts into bulk carriers. As the crisis unfolded, shipbuilders even managed to secure contracts for the construction of new dry bulk tonnage, despite the surplus carrying capacity in this sector.

The severe problems in the tanker market have often been used to explain the negative development in the dry bulk sector.[4] Combination carriers provide the most direct link between the two. Shortly before the shipping crisis, eighty percent of combination carrier tonnage was employed in the tanker market due to the strong demand for oil transport. After the tanker market broke down, its share of combination carrier employment fell to fifty-eight percent in 1974 and to fifty-one percent by 1975. This meant that combination carrier tonnage offered in the dry bulk market increased. Whereas dry cargoes had only constituted about fifteen percent of the volume of goods transported by combination carriers in 1972, this increased to nearly half by 1975.

Two features explain why as much as half the combination carrier fleet was still engaged in the depressed tanker market in 1975. First, the crisis had a considerable impact on the dry bulk market as well. Second, approximately twenty-eight percent of the total combination carrier fleet was commit-

---

[3]H.P. Drewry Shipping Consultants Ltd., *Shipping Statistics and Economics*, No. 44 (June 1974), 5.

[4]For examples of the use of this causality, refer to Norway, Stortingsmelding No. 52 (1980-1981), "Om skipsfartsnæringen" ["On the Shipping Industry"], 16; and Michael Beenstock and Andreas Vergottis, *Econometric Modelling of World Shipping* (London, 1993), 37.

ted to oil on period charters.[5] Moreover, the hardest hit tankers were larger than many of the combination carriers offered in this market. Still, the influx of combination carriers had a considerable impact on the dry bulk market.

In 1972 three million deadweight tons (dwt) of combination carriers were employed in the dry bulk sector. Three years later this had increased almost sevenfold to more than twenty million dwt.[6] Combination carriers comprised less than five percent of total transport capacity in the dry bulk market in 1972, but by 1975 it was seventeen percent. By 1981 more than three-quarters of the volume transported by combination carriers consisted of dry bulk goods.

As a result of the redeployment of vessels and net growth in the combination carrier fleet until 1979, the volume of tonnage in the dry bulk market mushroomed. In the early 1980s almost thirty-five million dwt of combination carrier tonnage was employed in the dry bulk sector. The bulk carrier fleet at the time amounted to some 142 million dwt, so the combination carriers actually contributed to a twenty-five percent increase in supply.

The combination carriers did not switch from the tanker to the dry bulk market immediately after the former collapsed (see figure 4.1). Dry bulk companies earned reasonable profits in the first half of 1974, mainly due to continuing growth in the trade of important commodities. But by the summer of 1974, the dry bulk sector had started to feel the effects of the oil price increase. Combination carriers exerted a downward pressure on bulk freight rates. Initially, rates for large vessels fell the most, but by the end of 1974 dry bulk rates had plunged to a level at which it was no longer profitable for the largest combination carriers to switch from the tanker to the dry bulk market.[7] The changing employment of combination carriers contributed to an increase in the breadth of the international shipping crisis by affecting even those owners who had refrained from investing in tankers.

The influx of combination carriers showed dry bulk owners what they were up against. As the tanker crisis progressed, competition intensified. The contracting boom before the freight market collapsed in the autumn of 1973

[5]OECD, *Maritime Transport* (Paris, 1976), paragraph 126. This figure was expected to fall to less than ten percent by the end of 1978.

[6]Fearnley and Egers Chartering Co. Ltd., *Review 1976* (Oslo, 1977), 15 and 21. See figure 4.1 for information on how tonnage has been measured. Ore/oil ships accounted for slightly more than half of the combination carrier fleet. Seventy-five percent of the fleet comprised ships larger than 80,000 deadweight tons (dwt), and almost sixty percent of the fleet was accounted for by vessels of more than 100,000 dwt; see "Markedet for kombinerte skip" ["The Market for Combination Carriers"], *Kapital*, No. 11 (1973), 22.

[7]Fearnley and Egers Chartering Co. Ltd., *Review 1974* (Oslo, 1975), 33.

meant that shipyard order books were thicker than ever before. The world tanker order book on 1 January 1974 amounted to 193 dwt, almost ten times as much as eight years earlier. The orders for combination and dry bulk carriers were ten and twenty-three million dwt, respectively. By the end of 1974, several shipowners had realised that there would be a persistent oversupply of tanker tonnage and consequently wanted to avoid the delivery of vessels that might have to be laid-up immediately after delivery.

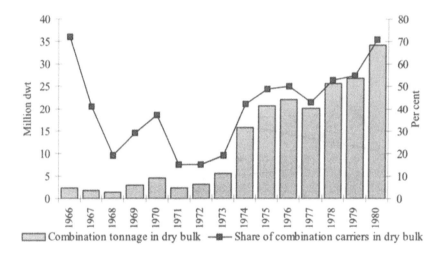

Figure 4.1:   Employment of Combination Carriers, 1966-1980

Note: The amount of combination tonnage operating in the dry bulk market, measured in dwt and represented on the left axis, has been found by multiplying the share of combination carrier activity in the dry bulk market by the size of the combination carrier fleet. The figure for 1978 may be exaggerated due the inclusion of inactive tonnage. The right axis refers to the share of combination carriers operating in the dry bulk market as a percentage of the total combination carrier fleet.

Source: Fearnley and Egers Chartering Co. Ltd., *Review 1972* (Oslo, 1973); and *Review 1981* (Oslo, 1982), tables 3 and 15.

One way that shipowners could avoid the delivery of superfluous tanker tonnage was by cancelling existing orders, and in several instances it was in the interest of both the shipowner and the shipyard to void a contract. Several yards had not foreseen the enormous increase in building costs when accepting orders and were therefore willing to cancel contracts that had become unprofitable. In some instances shipowners were willing to pay a penalty to get out of unprofitable contracts in order to secure their financial position.[8]

Conversion was often a less expensive alternative to cancellation.[9] The tanker contracts that were converted to combination or bulk carriers were delivered after 1976 into a dry bulk market which by then was also experiencing a tonnage surplus. The conversion of tanker contracts thus contributed to the spread of the shipping crisis to the dry bulk sector and to a lengthening of the crisis in that sector.

Figure 4.2: Dry Bulk and Combination Tonnage: Percent of Total Contracting, 1963-1980

Source: Fearnley and Eger, *Review 1972*; and *Review 1980* (Oslo, 1981), table 10.

When activity in shipyards is slack, governments are more willing to use subsidies to secure contracts. After the tankers ordered during the boom had been cancelled or delivered, shipbuilding capacity was no longer in short supply. Shipyards had to look elsewhere for business, as it was almost impos-

---

[8]Helge W. Nordvik, "Norwegian Maritime Historical Research during the Past Twenty Years: A Critical Survey," *Sjøfartshistorisk Årbok 1990* [*Norwegian Yearbook of Maritime History 1990*] (Bergen, 1991), 253.

[9]A survey of contracting activity in 1974 claims that most of the converted tanker contracts were substituted by ore/bulk/oil (OBO) or bulk carrier orders, particularly for Panamax-size vessels; see "Konverteringenes år" ["The Year of Conversions"], *Norwegian Shipping News*, No. 10 (1975), 44.

sible to build tankers at prices that were acceptable to them and profitable for investors as well. The subsidies and the shipyard capacity were consequently allocated to the relatively functional parts of the shipping market. As it was no longer profitable to build tankers – and because the yards could use subsidies as bait – the available construction capacity was used to build combination and bulk carriers, as well as more specialised vessels (see figure 4.2).[10] The long-term result of the conversion of tanker contracts and the increased construction of combination carriers and dry bulk tonnage was that the supply side in the dry bulk market continued to expand, despite the market constraints. But this was not the only reason for the deterioration of conditions in the dry bulk market, for the demand side aggravated an already difficult situation.

In Fearnley and Eger's *Review 1975*, four factors were highlighted to explain the difficulties in the dry bulk market.[11] The first was the negative impact of the tanker market collapse. In addition, reduced steel production, conditions in the international grain trade and congestion in major ports were used to explain market conditions.[12]

The demand for bulk transport capacity is closely related to global manufacturing activity, and the industrialised countries had entered a recession by the mid-1970s (see table 4.2). Despite the fact that industrial production and world trade increased in the 1970s, growth was weaker than in previous decades and far below expectations. At the same time, the international monetary system was radically changed, and the industrialised countries had to combat the twin evils of inflationary pressure and high unemployment. The focus on inflation reduction, exemplified by deflationary monetary policies and relatively tight fiscal controls, aggravated the recessionary forces.

In an analysis of the demand for bulk transport between 1965 and 1975, Siri Pettersen Strandenes found that fluctuations in the actual and potential Gross National Product (GNP) among OECD nations explained a large portion of the variations in demand for seaborne transport. The difference be-

---

[10]See, for example, Peter Mottershead, "Shipbuilding: Adjustment Led Intervention or Intervention Led Adjustment," in Geoffrey Shepherd, François Auchen and Christopher Saunders (eds.), *Europe's Industries: Public and Private Strategies for Change* (London, 1983), for a survey of the activity in world shipbuilding from 1966 until the end of the 1970s. In addition to dry bulk vessels, ships intended for other market segments represented alternative employment for shipbuilders.

[11]Fearnley and Egers Chartering Co. Ltd., *Review 1975* (Oslo, 1976), 8.

[12]The development of grain transports exercised a positive influence on the demand for dry bulk transport capacity in 1975. This shows that one of the most important dry bulk commodities developed independently of global industrial activity. Moreover, congestion may also exert a positive effect on the demand for tonnage.

tween actual and potential GNP can be seen as a measure of unused capacity.[13] Reduced demand for transport might therefore be explained by variations in the production in the industrial countries and in particular by the business cycle development in some key sectors.

Table 4.2
Annual Growth of Total Production, Decadal Averages, 1950-1980

	1950-1959	1960-1969	1970-1979
Japan	9.5	10.5	4.9
West Germany	7.9	4.8	2.8
Great Britain	2.7	2.8	1.8
United States	3.2	4.3	3.0

Note: Japan, which due to its geographic position and need for imported raw mate-rials, is particularly important in the demand for shipping services, experi-enced the largest relative decrease in production growth. When averages de-noting the growth in the various decades are used, the high economic growth in the beginning of the 1970s partly cloaks the differences between economic growth in the 1960s and the 1970s.

Source: Herman van der Wee, *Prosperity and Upheaval. The World Economy 1945-1980* (London, 1986), 50.

The recession began in Japan at the beginning of 1974 and was evi-dent in the US a few months later.[14] In late 1974 European manufacturing ac-tivity began to decline as well. Despite the recession and a reduction in steel production, dry bulk demand increased in 1974.[15] But the next year conditions were sufficiently bleak that demand actually fell. Steel, traditionally quite im-portant for dry bulk demand, was particularly hard hit by the economic slump. It has been estimated that steelworks achieved only about sixty or seventy per-cent utilisation in 1975. The result was a large reduction in the demand for iron ore and coal, the two most important dry bulk commodities (see figure 4.3). Yet while the seaborne transport of iron ore fell by almost seven percent

---

[13]Siri Pettersen Strandenes, "Trekk ved konjunkturutviklingen i tankfarten" ["Aspects of Business Cycle Development in Tanker Shipping"] (Unpublished høyere avdeling thesis, Norwegian School of Economics and Business Administration, 1977). A similar study of the dry bulk sector is referred to in Victor D. Norman, *The Econom-ics of Bulk Shipping* (Bergen, 1979), 6.

[14]Johan Seland, *Norsk skipsfart år for år 1946-1976 [Norwegian Shipping Year by Year, 1946-1976]* (Bergen, 1994), 235.

[15]In part because steelworks were unable to adapt immediately to price changes in energy and transport, there was a growth in the carriage of iron ore and coal in 1974.

in 1975, the carriage of coal actually rose by more than eleven percent due to its increased used for energy purposes outside the steel sector.

The dry bulk market decreased in 1975 for the first time since the end of the 1950s because the increase in the transport of grain and coal was insufficient to neutralise the negative influence of reduced economic activity in general.[16] The seaborne transport of iron ore fell every year from 1975 until 1978; in the latter year it was about twelve percent lower than in 1974. The demand for coal transport was unstable, but on average twenty percent above the level of the pre-crisis years.

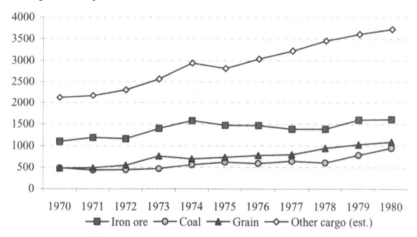

Figure 4.3: World Seaborne Trade of Non-Liquid Goods, 1970-1980 (billion ton-miles)

Source: Fearnley and Eger, *Review 1980*, table 2.

Due to the oil price increase the industrial world tried to reduce its need for energy or to meet it by using sources other than petroleum. Oil's share of OECD primary energy consumption was reduced from 53.4 percent to 45.4 percent between 1973 and 1982, while the share of coal rose from 18.3 to 21.4 percent. Energy-saving measures were important as well. The total amount of energy consumed, which had increased heavily throughout the post-war era, was lower in 1982 than it had been nine years earlier.[17]

---

[16]Some of the increased demand for grain transport was satisfied by tankers. According to calculations based on figures from Fearnley and Egers Chartering Co. Ltd., *Review 1977* (Oslo, 1978), tables 1 and 6, approximately five percent of grain shipments was undertaken by tankers.

[17]British Petroleum, *BP Statistical Review of World Energy 1984* (London, 1984), 27-28. The share of nuclear power increased from 1.2 to 5.3 percent over the

The spread of the crisis from the tanker sector was only one of the short-term effects of the shipping crisis. The transformation initially became evident through the development of tanker freight rates, but shortly thereafter the value of vessels and capacity utilisation also reflected the new conditions.

The strong fluctuations in spot market freight rates indicate that the price-making process was extremely flexible (see figure 4.4). As early as one month after OPEC had announced a reduction of the oil supply, rates on some of the spot market voyages had fallen by an astonishing eighty-eight percent.[18] Spot tanker rates declined first and most violently and remained low until a moderate resurgence in the summer of 1979. This rate increase was the result of the tonnage shortage that came about due to the high price of bunkers, which facilitated slow steaming and other types of inefficiencies in the tanker fleet.[19] It affected mainly tankers under 100,000 dwt. The short-lived increase in freight rates was followed by a recession as serious as the one in the 1970s. In the early 1980s, the tanker market was in a worse state than ever.

Figure 4.4 shows the trends in spot market freight rates in the 1970s. The rate level was not necessarily a good indication of profitability and was not perfectly correlated with the economic performance of shipowners. The main reason for this was the massive increase in the cost of bunkers, which implied that Worldscale (W) on various occasions was insufficiently adjusted to account for shifting costs. Before the oil price increase, W 100 was US $10 per ton of cargo on trips from the Persian Gulf to Europe. In 1975 W 100 constituted US $14.50 per ton, an increase of forty-five percent.[20] In the period after 1974, fuel costs accounted for the majority of the operating costs of tank-

---

same period. The figures from 1982 include the effect of OPEC II, which prompted the introduction of further energy-saving measures. In the period 1973-1979 total consumption of primary energy in Western Europe, North America and Japan increased by approximately ten percent.

[18]Arnljot Strømme Svendsen, "Skipsfartskonjunkturene i 1970-årene" ["Shipping Cycles in the 1970s"], *Sjøfartshistorisk Årbok 1978* [*Norwegian Yearbook of Maritime History 1978*] (Bergen, 1979), 228. Calculations from the aforementioned example of rates were at Worldscale (W) 475 and W 55. The variations in figure 4.4 are not as violent because the figure is based on monthly registrations of weekly averages. The example cited in Strømme Svendsen's article is based on extreme observations, and the difference will therefore be larger than for weekly or monthly averages.

[19]Beenstock and Vergottis, *Econometric Modelling*, 62.

[20]Worldscale rate levels in different years are thus not comparable because there was a lag in costs in years with large changes in the cost of bunkers. It is also necessary to take into account exchange rate fluctuations, which were more violent in the 1970s than previously.

ers and bulk carriers, so there is little doubt that changes in the cost of bunkers were important for explaining fluctuations in shipping profitability.[21]

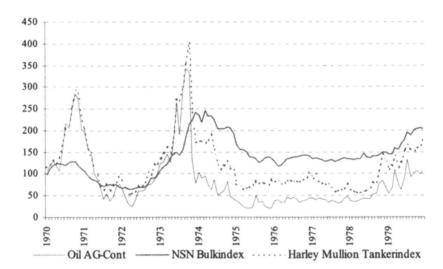

Figure 4.4   Monthly Freight Rates in the Spot Market, 1970-1979 (Worldscale)

Source: Fearnley and Eger, *Review*, various years.

One important aspect of the pattern of freight rates was the increasing difference between rates for ships of various sizes. Freight rates for the largest vessels, the Very Large Crude Carriers (VLCCs) and Ultra Large Crude Carriers (ULCCs), had fallen to particularly low levels. One reason for the increased variation was the difference in supply and demand for the various size classes. Before 1973 fleet growth had been highest among the largest vessels. But the dislocations caused by the crisis favoured smaller ships, which were far more flexible with regard to the variety of liquid goods they could transport. Supertankers, on the other hand, were purpose-built to transport crude oil across long distances and were of limited use outside this market.[22]

The freight rate decline in the dry bulk market was neither as pronounced nor as rapid as with tankers. The crisis did not have a serious impact on dry bulk rates until late 1974. Again, there were large differences between various ship classes. In 1974 rates were still satisfactory for smaller vessels,

---

[21]Michael Beenstock and Andreas Vergottis, "An Econometric Model of the World Market for Dry Cargo Freight and Shipping," *Applied Economics*, XXI, No. 3 (1989), 52.

[22]See David Glen, "The Emergence of Differentiation in the Oil Tanker Market, 1970-1978," *Maritime Policy and Management*, XVII, No. 4 (1990), 289-312.

while for those over 70,000 dwt rates were more similar to tankers. Low rates spread, however, and in 1975 only ships under 40,000 dwt earned acceptable rates. Dry bulk rates remained low until the temporary freight rate increase in 1979. The demand for bulk transport then grew faster than the fleet, thus reducing surplus bulk carrier capacity.

Like the close relationship between freight rates and contracting, there was a high correlation between rates and ship prices.[23] The freight market collapse in the autumn of 1973 initially caused an abrupt halt in the sale of used tankers, with Fearnley and Eger commenting that "buyers [in the second-hand tanker market] were equally cool as they were hot at the same time last year." During 1973 second-hand values of tankers increased strongly. The value of a 100,000-dwt vessel built in 1967-1968 was estimated at $13.5 million at the start of the year, whereas the value in October was $30 million.[24]

The effects of the strong increase in the value of tankers in 1973 were neutralised the next year, and activity in the second-hand tanker market was minimal in 1974. The considerable shift in vessel prices between the late 1960s/early 1970s and the period after the collapse is evident when figure 4.5 is compared with figure 3.20. Along with the decline in values, there was a change in the relative price of turbine and motor tankers, as the value of the former declined far more than the latter.[25] A turbine-driven VLCC, which in 1972 had been twice as expensive as a 100,000-dwt motor tanker, had the same value as the far smaller ship a few years later.

According to S.R. Tolofari, there "is no doubt that the type of engine will play an important role in deciding what rate a shipowner would accept and a charterer pay, in view of the enormous fuel consumption differential between turbine and diesel tankers."[26] Shipowners operating turbine tankers were adversely affected by the oil price increase, losing competitiveness when they were unable to compete with the more economical diesel ships. Another reason

---

[23]Most models that depict the shipping market contain assessments of the relationship between shipping and shipbuilding. One example of a survey where this relationship is the main focus is W. Charemza and M. Gronicki, "An Econometric Model of World Shipping and Shipbuilding," *Maritime Policy and Management*, VIII, No. 1 (1981), 21-30.

[24]Fearnley and Egers Chartering Co. Ltd., *Review 1973* (Oslo, 1974), table 17. The ship prices are meant to be representative of the value of a given type of vessel and are calculated on the basis of current transactions in the second-hand market.

[25]In Fearnley and Eger, *Review*, various years, there is no comparison of the price development for motor or turbine tankers in the same size class because most of the turbine-driven vessels were large tankers.

[26]S.R. Tolofari, "Open Registry Costs and Freight Rates: Are They Related?" *International Journal of Transport Economics*, XIV, No. 1 (1987), 90.

for the relative decline in the value of turbine tankers was that this technology had primarily been used on the largest vessels, the ones that were faced with the most adverse conditions.

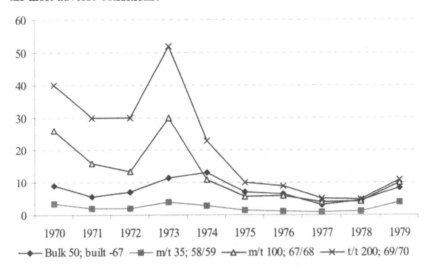

Figure 4.5: Average Values of Second-hand Vessels, 1970-1979 (US $ million)

Note: The prices are market value estimates at valid exchange rates for charter-less vessels in good condition and with fairly prompt delivery on a cash basis. The legends indicate the type of ship (bulk, motor tanker or turbine tanker), followed by vessel size in '000 dwt and year of construction.

Source: Fearnley and Eger, *Review*, various years, tables 18 and 19.

The depressed freight rates were felt in the market for new buildings as well. The price of new VLCCs and ULCCs fell by approximately twenty percent during 1974.[27] Since new building prices are relatively rigid – whereas there is an extremely large degree of flexibility in the second-hand market – the violent fluctuations in the market for used vessels did not occur to the same extent in the market for new vessels.[28] Moreover, second-hand prices tend to adapt to large exogenous shocks, such as the oil price increase, relatively rapidly, whereas it takes time before the price of new buildings adjusts to new

---

[27]Fearnley and Egers Chartering Co. Ltd., *Review 1974* (Oslo, 1975), 38.

[28]Michael Beenstock, *A Theory of Ship Prices* (London, 1984), 9.

conditions.[29] Material and labour costs represent a floor through which new building costs, at least in theory, should not fall. The shipbuilding industry, however, does not necessarily conform to such common economic principles. Due to increased subsidisation by governments, new vessels were sold at prices which failed to reflect the actual cost of construction.

The fall in the value of ships did not appear at the same time in all markets. Again, the response in the dry bulk market was slower and less violent than the tanker experience. In 1974 the price of some dry bulk carriers continued to increase, partly as a result of the small amount of modern tonnage available following the tanker boom. Since global shipbuilding capacity is limited, the focus on tanker construction in the early 1970s resulted in limited new building activity for the dry bulk market. The strong growth in the cost of building ships contributed to the increase in the price of second-hand dry bulk tonnage in 1974.

The reduction in the value of the world fleet did not have a uniform effect on shipping companies. A reduction in the value of a company's vessels implies a deterioration of the firm's equity. Thus, the assets of those companies that had invested largely in the biggest turbine-driven tankers developed in a particularly negative manner. Moreover, since these were generally both relatively new and relatively expensive, only a small share of the loan principal had been repaid. The large loans that usually were necessary to finance these vessels meant that the financing costs were particularly high.

Due to the surplus capacity, a large share of the world fleet had to be laid-up, find alternative forms of employment or run at speeds far slower than before the crisis. The large lay-ups were perhaps the most conspicuous feature of the crisis, particularly in Norway, where a considerable number of large tankers was mothballed in the fjords. Accordingly, the effects of the crisis became glaringly visible for other than industry insiders. Several attempts were made to find employment, other than transport, for the ships, and in some countries vessels were used for storage.[30] In addition, the effective transport capacity was reduced sharply because it was no longer economical to operate the vessels at full speed.

---

[29]According to Beenstock and Vergottis, *Econometric Modelling*, 62, "[o]vernight, VLCCs values fell by [US] \$20 million." A very interesting graphic representation of the influence of freight rates on the second-hand value of tankers is presented in S.R. Tolofari, K.J. Button and D.E. Pitfield, "The Cost Structure of the Tanker Sector," *International Journal of Transport Economics*, XIV, No. 1 (1987), 76.

[30]One imaginative suggestion was the use of tankers as a source of electricity; see "Use of Tankers as Auxiliary Electricity Suppliers Can Absorb Excess Tonnage," *Norwegian Shipping News*, No. 19 (1978), 8-10, where it is claimed that the tonnage laid-up in Norwegian fjords at the start of 1978 could supply thirteen percent of the nation's demand for electricity.

Fearnley and Eger has estimated the tonnage overcapacity in the international market for shipping services for the years between 1974 and 1978, using 1973 as a base year since there was at least a notional equilibrium between tonnage supply and transport demand (see figure 4.6).[31] With this as a starting point, the company estimated the tonnage surplus during the crisis and calculated how the excess tonnage was absorbed (see table 4.3).[32]

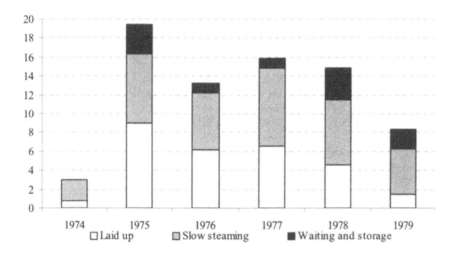

Figure 4.6:   Estimated Tonnage Surplus as Share of Total Fleet, 1974-1979

Note: Corresponding figures for surplus tankers would be considerably higher.

Source: Fearnley and Eger, *Review*, various years.

The figures show that a considerable share of the tonnage surplus was absorbed through mechanisms other than lay-ups. It is interesting to note that in most years during the crisis, slow steaming, the effects of which are sometimes neglected in a discussion of the crisis, was more important than lay-ups in reducing the surplus. Slow steaming is undertaken when the fuel cost savings are higher than the profits foregone in additional voyages. Due to the in-

---

[31]Despite the fact that 1,000,000 dwt were laid-up or seeking cargoes at the end of the year, it is correct to regard 1973 as a year with a reasonable balance between demand and supply.

[32]The calculations in table 4.3 and figure 4.6 are estimates based on the assessments of the tonnage balance in Fearnley and Eger, *Review*, various years. These assessments were not always internally consistent, but the estimates can nevertheless indicate the manner through which the tonnage surplus was expressed.

creased price of bunkers, the optimal operating speed of the vessels became
three or four knots slower than before. The reason was that the use of bunkers
to achieve an increase in speed increased disproportionately. Optimal speed is
thus inversely correlated with the price of bunkers.

Table 4.3
Tonnage Balance at the End of the Year, 1973-1979 (million dwt)

	World Fleet	Tonnage Surplus	Slow Steaming	Laid-up	Waiting and Storage
1973	439.9	1		1	
1974	493.9	15	11.5	3.5	
1975	543.7	105.7	40	48.7	17
1976	591.3	78.3	36	36.3	6
1977	624.6	99	52	41	6
1978	632.7	96.9	44	28.9	21
1979	637.9	52.8	30	9.8	13

Note: Figures refer to 1 January of the following year.

Source: Fearnley and Eger, *Review*, various years.

Table 4.4
Fuel Consumption of a 270,000-dwt Oil Tanker
Fuel Consumption (tons)

Speed (knots)	Per 24 Hours at Sea	Annual	Avg. Fuel Cost (million $)	Annual Transport (million tons)	Fuel Costs per Ton Transported
11	80	26,000	2.3	1	2.3
14	125	41,000	3.7	1.3	2.8
16	170	56,000	5.5	1.5	3.7

Source: Yvan Du Jonchay, *The Handbook of World Transport* (London, 1978), 11.

The fact that freight rates had plummeted also made it increasingly
profitable to reduce the speed of a ship (see table 4.4). Shipping companies
chose to let their ships sail at lower speeds as a result of the reduced availability of cargoes and bunkers. Charterers, such as oil companies, also opted for
slow steaming when the savings in bunker costs by the in-house and chartered
fleets were higher than the cost of obtaining additional transport capacity in the
spot market. Although the low freight rates implied that the cost of buying
additional capacity was low, the high price of bunkers made a reduction of the
speed of both company-owned and chartered tonnage profitable. During the
shipping crisis, speeds of less than ten knots were common.

    As a result of the increase in the price of bunkers and the difficulties
in securing employment, there was a permanent reduction in the utilisation
coefficient of the vessels, meaning that the annual transport undertaken by each

vessel was lowered. In aggregate, this implies that there was a reduction in the effective tonnage, i.e., the amount of cargo which the world fleet was able to transport.[33] Effective tonnage was reduced when ships sailed at lower speeds, were used for storage or were fitted with segregated ballast tanks.[34]

Figure 4.7 shows the sharp increase in the lay-up rate. The figure is based on estimates of the world fleet and tonnage laid-up measured in dwt and only takes into account tankers, bulk carriers and combination carriers above 10,000 dwt. Consequently, the lay-up rate might differ from surveys based on gross registered tons (grt) or lay-ups as share of the total fleet, including liners and passenger vessels.

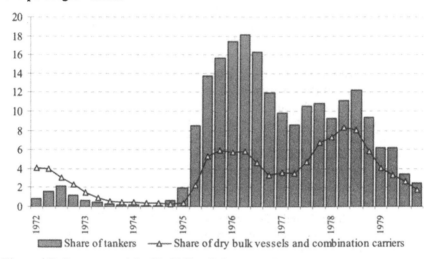

Figure 4.7: Percentage of the World Fleet Laid-up, 1972-1979 (quarterly)

Source: Fearnley and Eger, *Review*, various years, tables 3 and 7.

The advantage of laying-up ships is the reduction in operating costs. But financing costs still accrue, and there are also costs specifically connected to preparing ships for lay-up. Expenses to lay-up a VLCC might be as much as $75,000 a month, in addition to a one-off cost of preparation and site acquisition in the region of $700,000.[35] An alternative was to use vessels for storage,

---

[33]This was more than offset, however, by the increase in real tonnage in the first years after the freight market collapse.

[34]See P.G. Bennett, C.S. Huxham, and M.R. Dando, "Shipping in Crisis: A Trial Run for 'Live' Application of the Hypergame Approach," *Omega*, IX, No. 6 (1981), 584.

[35]OECD, *Maritime Transport 1975*, paragraph 221.

sometimes as part of a transport assignment. The low alternative value of the tonnage in a market affected by overcapacity resulted in a situation where it became economical to use them for other purposes than merely transport.

In some instances, the oil companies offered to allow shipowners to let their ships "wait" outside the land-based terminals before delivering the cargo. This meant that the oil companies used the tankers in a strategic manner, speculating in oil price changes and supplementing the storage capacity of their refineries by means of tankers. In 1978, five million dwt of tanker capacity was used for storage purposes off the coast of Japan.[36]

Another reason that tankers were positioned off the coast was the transhipment of oil from large tankers with drafts too deep for the ports to smaller tankers which could distribute the oil to terminals ashore. Following the growing use of large tankers, the level of port constraints increased in the period 1973-1979, indicating that a higher share of tanker tonnage was unable to use the most important ports or terminals if fully loaded.[37] Deliveries from the Middle East to the US were often organised in this manner, as there were no port facilities which could cater for the largest vessels. In 1975 three Canadian (but no American) ports were capable of taking full- or part-loaded VLCCs.[38] The largest US ports, Richmond and Ferndale, could only accommodate ships under 150,000 dwt in a period when most of the long-range American oil imports were transported on VLCCs and ULCCs. Ships were also left idle while they were awaiting employment. These vessels would not be considered as laid-up, and at times as many as fifty tankers were awaiting cargoes off Kharg Island. This area was sarcastically nicknamed "Reksten Bay" due to the large number of ships that the Norwegian shipowner Hilmar Reksten had waiting here.

Despite the large temporary imbalances in the shipping market, some mechanisms contributed towards an improvement of the tonnage balance. One is scrapping. When freight rates and second-hand prices fall, an increase in the demolition of ineffective tonnage usually follows. From the second half of the 1970s the adverse conditions contributed to an increase in the number of vessels leaving the fleet. Even though breaking prices were fairly stable throughout the latter part of the decade, the fall in second-hand values made scrapping relatively profitable.

Figure 4.8 shows the strong increase in the scrapping of tanker tonnage as the shipping crisis progressed. One important aspect was the increase in the average size of tankers scrapped, which grew from approximately 7500

[36]Fearnley and Egers Chartering Co. Ltd., *Review 1978* (Olso, 1979), 38.

[37]Glen, "Emergence," 297-298.

[38]H.P. Drewry Shipping Consultants Ltd., *The Trading Outlook for Very Large Tankers* (London, 1975), 34.

grt in 1972 to 24,000 grt in 1980. In 1977 the largest vessel broken up was 170,000 dwt, whereas in 1978 several VLCCs in good condition were sold to breakers.[39] The average size of tankers scrapped in figure 4.8 is relatively low due to the inclusion of gas and chemical tankers in the figures.

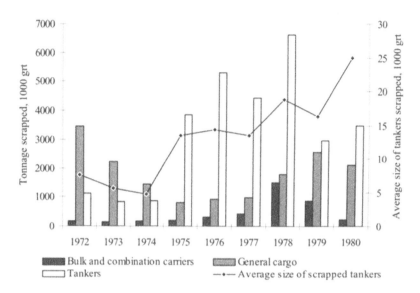

Figure 4.8: Scrapping by Vessel Type and Average Tanker Size, 1972-1980 ('000 grt)

Note: The tanker share of total scrapping would have been more prominent, and general cargo less so, if dwt had been used as the basis for comparison. The figures correspond with similar data from Fearnley and Eger, *Review*: 6.6 million grt of tanker tonnage and 1.8 million grt of general cargo tonnage were broken up in 1978. This corresponds to 13.5 million dwt and 2.6 million dwt, respectively.

Source: OECD, *Maritime Transport*, various years.

The most important effect of the oil price increase initially was the drastic reduction of tanker freight rates, particularly in the spot market. As the crisis escalated, other segments were affected by the dire market conditions, and there was an increase in the difference between rates for ships of different types and size classes. Through the influx of combination carriers, the conversion of new building orders and the shipyards' desperate measures to sustain activities, the tanker crisis evolved into a more general shipping crisis. The difficulties were aggravated by the recession in the industrialised countries.

---

[39]Fearnley and Eger, *Review 1978*, 42.

Chapter 5
Stage Two of the Crisis and the Long-term Changes

By the mid-1970s the problems in the tanker market had evolved into a full-fledged shipping crisis, and no part of the shipping sector was unaffected by the calamities. But at the end of 1978 the prospects suddenly looked much brighter, and it seemed as though the problems were nearing an end. Yet this proved to be just a temporary reprieve, and in the first half of the 1980s the situation deteriorated even further.

The shipping crisis implied drastic changes in freight rates, vessel values and capacity utilisation. When the mismatch between supply and demand was temporarily reduced near the end of the 1970s, the markets responded quickly, leading to a rapid increase in freight rates, ship values, contracting and capacity utilisation. In the period from March to December 1979, tanker lay-up rates were reduced by approximately two-thirds. The spot market for Very Large Crude Carriers (VLCCs) and Ultra Large Crude Carriers (ULCCs) rose from Worldscale (W) twenty-two in February to a peak at W seventy-one five months later, and the value of the largest class of tankers almost doubled between 1978 and 1979.[1] Moreover, the volume of tanker tonnage ordered increased from three million deadweight tons (dwt) in 1978 to more than fourteen million dwt the following year.

Although the improved conditions around the turn of the decade were merely a false recovery, it makes sense to analyse the two stages of the shipping crisis independently, since the basis for the second stage differed from the first. In the 1970s the problems were caused by lower-than-expected demand growth and a massive tonnage increase. Stickiness caused by the length of time between ordering and delivery and the high cost of cancellations meant that the fleet grew while demand stagnated.

In the first half of the 1980s there was an absolute reduction in shipping demand. Even a reduction in the tanker fleet was insufficient to neutralise the negative effects. The basis was an absolute decline in oil consumption and transport, following the strong increase in oil prices in 1979/1980. The move-

---

[1]Lay-up figures are from Fearnley and Egers Chartering Co. Ltd., *Review 1979* (Oslo, 1980), table 7; freight rate figures are from *ibid.*, table 13, and ship values are based on the value estimation of a 350,000-deadweight ton (dwt) turbine tanker, estimated at sixteen million US dollars in 1977, twenty-two million in 1978 and forty million in 1979; see Fearnley and Egers Chartering Co. Ltd., *Review 1981* (Oslo, 1982), table 19.

ment of crude oil in 1985 was less than forty percent of the volume in the peak
year of 1977.[2]

Figure 5.1 shows an indexed representation of developments with
1970 – a year with relative balance between demand and supply – used as the
base year. The strong growth before 1973 is evident, as is the stagnant demand
and growing fleet from 1974 onwards. During the second stage of the crisis,
fleet development played a less prominent role. The world tanker fleet was
reduced after 1978, and the increased scrapping from 1982 actually contributed
positively to a reduction of the potential tonnage surplus. But at the heart of
the problems lay a significant reduction in shipping demand. The fall in demand for tanker transport was so large that even a decrease in the world fleet
was insufficient to bring about anything resembling balance between supply
and demand. By 1982 the demand for tanker shipping was slightly lower than
it had been in 1970, but over the same period the tanker fleet had increased by
almost 150 percent.

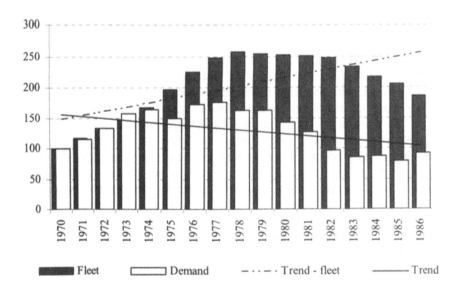

Figure 5.1:   Tanker Demand and Supply, 1970-1986 (1970=100)

Source: Fearnley and Egers Chartering Co. Ltd., *Review*, various years.

---

[2]There was a small increase in the demand for the transport of oil products
between 1977 and 1985. Nevertheless, even when this modest increase is taken into
account, total seaborne oil movements were more than halved over the period.

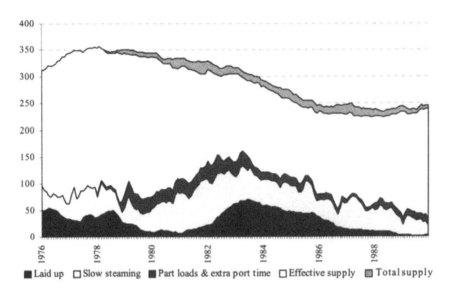

■ Laid up   □ Slow steaming   ■ Part loads & extra port time   □ Effective supply   ▤ Total supply

Figure 5.2: Tanker Tonnage Imbalance, 1976-1989 (million dwt)

Note: Effective supply is measured as total tanker and combination carrier tonnage available minus tonnage used for storage. At the peak in late 1981, twenty-two million dwt were used for storage, implying that effective supply was almost seven percent lower than actual supply.

Source: Monthly figures from Gilbert Jenkins, Martin Stopford and Cliff Tyler, *The Clarkson Oil Tanker Databook* (London, 1993), tables III-12 to III-25.

Figure 5.2 reveals the increasing imbalance between tanker demand and supply at the beginning of the 1980s. In April 1983, the nadir of the crisis, the total supply of tankers and combination carriers was 310 million dwt, whereas basic demand was 137 million dwt. The actual surplus thus amounted to 173 million dwt, indicating an overcapacity of more than 125 percent. This implies that more than two vessels were chasing every available cargo of oil. Approximately one-quarter of the surplus was reduced through the use of tonnage for storage, the acceptance of part cargoes and extra waiting time in ports. Of the effective surplus of 134 million dwt, more than a quarter was laid-up and the rest neutralised through slow steaming. The fifty percent decline in the demand for crude transports from 1980 to 1985 meant that the total demand for seaborne trade, measured in ton-miles, was reduced by approximately twenty-five percent.[3] The drastic changes in the tanker market follow-

---

[3]Calculated from Fearnley and Egers Chartering Co. Ltd., *Review 1986* (Oslo, 1987), 30.

ing the two peaks at the beginning of the 1970s, and the continuation of the crisis into the 1980s, are clearly illustrated by the development of freight rates (see figure 5.3). Although the amount of tanker tonnage was reduced after 1978, freight rates did not recover properly.

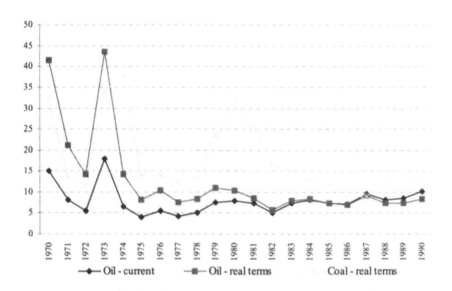

Figure 5.3: Freight Rates in Current and Real Terms, 1970-1990 (US $ per ton)

Note: Figures have been deflated by the American wholesale price index, with 1985 as the base year. The graphs depicting oil are spot rates from the Middle East to Northwest Europe or the US east coast. The graphs depicting coal refer to a 50,000-dwt vessel on the route from Hampton Roads to Japan. As the graph is based on averages, some fluctuations are concealed. For instance, oil rates increased from US $3.5 per ton in January 1987 to almost US $11 per ton in the middle of the year.

Source: Annual averages from Jenkins, Stopford and Tyler, *Clarkson Oil Tanker Databook*; and Fearnley and Eger, *Review*, various years.

The crisis left permanent marks on the shipping market. It is difficult to say to what extent it exacerbated, curtailed or altered trends in the 1970s and 1980s. Some features, such as the process of internationalisation, had been going on for decades, but the speed, if not the direction, of these developments might have been changed. The crisis also triggered adjustments in other sectors, particularly shipbuilding. Governments in several of the most important shipbuilding nations increased their influence as demand for new buildings fell. In the shipping industry there was a corresponding increase in government involvement, particularly in connection with the growing use of cargo reservation and financial assistance to local shipowners.[4]

As a result of the crisis, several companies, large and small, disappeared due to economic difficulties. The flag-of-convenience-fleets (FOCs) increased faster than the fleets of Traditional Maritime Nations (TMNs), indicating a greater separation of ownership from country of registration. Moreover, some new entrants acquired an important position in world shipping. At the international level, the structural development was characterised by the reduced importance of the TMNs. This transformation embodied two features. First, the increase of the FOC fleets continued, as the share of vessels registered in open registries such as Bahamas, Liberia and Panama grew from eighteen percent in the middle of 1970 to 27.5 percent in mid-1986. Second, developing countries increased their share of the world fleet from 7.7 to 20.3 percent over the same period.[5] The share of the world fleet registered in the FOCs or Asia thus doubled, from one-quarter in mid-1970 to half by the middle of the next decade. The Asian fleet increased especially fast, and this shift was particularly prominent in the first half of the 1980s, by which time the long-term effects of the freight market collapse had begun to take their toll. By 1987 the entire structure of the maritime industry had been transformed, as the hegemony of the TMNs, represented by the Organisation for Economic Cooperation and Development (OECD) registries, was successfully challenged by FOC countries and the Emerging Maritime Nations (EMNs, see figure 5.4).

For most of the postwar period the traditional industrial powers, represented by the members of the OECD, had controlled international shipping. Their position was reflected both in the standing of their domestic fleets and in the amount of tonnage they operated under FOCs. Indeed, the use of FOCs was largely an effect of intra-OECD competition. Although the FOC fleets to some extent were outside the jurisdiction of the TMNs, the companies with controlling interests generally were not.

---

[4]See *Norwegian Shipping News*, No. 21 (1977), 14.

[5]Gunnar K. Sletmo, "Shipping's Fourth Wave: Ship Management and Vernon's Trade Cycles," *Maritime Policy and Management*, XVI, No. 4 (1989), 299. These figures deviate from those presented later due to differences in the definition of flags of convenience. Due to such differences, the figures should be seen as indicative rather than absolute. It has been claimed that the term "flags of convenience" should be avoided because it is value-laden. Supporters of this institution prefer to use the term "flags of necessity." In the following analysis I have chosen to use the former term because it is much more common. This is a conscious choice on my part, but it is based on clarity of presentation rather than politics.

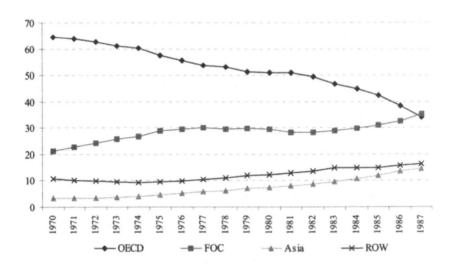

Figure 5.4: Distribution of the World Fleet, 1970-1987 (percent of dwt)

Note: Fleets of less than 400,000 gross registered tons (grt) are included in "Rest of the World." While there was some movement between the groups, this was not enough to affect the results significantly.

Source: OECD, *Maritime Transport*, various years.

The period from 1970 to 1987 was marked by an insignificant increase in the OECD fleet from 215.5 million to 218 million dwt. The latter figure was considerably lower than the OECD peak of 356 million dwt in 1978, and from 1970 to 1987 the OECD's share of the world fleet was almost halved, falling from 64.5 to thirty-four percent. The apparent reduction was partly mitigated by the transfer of OECD-owned tonnage to FOCs. The amount of tonnage registered under FOCs, on the other hand, increased considerably after the introduction of such registries in the interwar period, and their share of the world fleet more than doubled in the ten years after 1963 (see figure 5.5).[6] The FOCs have been controversial, and it has been claimed that they have become unregulated refuges for the fleets of shipowners with sub-standard vessels. This discussion, however, lies outside the scope of this

---

[6]FOCs existed before this – for instance, several British shipowners regarded Norway as an FOC in the late nineteenth century. But it was in the interwar period that some of the FOCs that came to dominate in the postwar period first rose to prominence.

book.[7] There have been attempts at the international level to regulate the registration of the world fleet through the introduction of the "genuine link" principle, among other methods. But the fleets of the FOC countries, and also the number of nations offering such registries, have increased considerably in the postwar period.

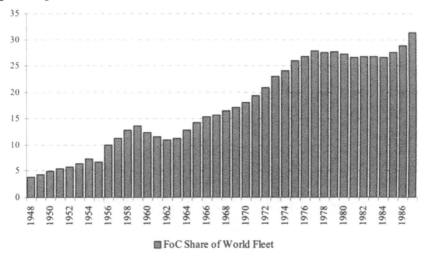

☐ FoC Share of World Fleet

Figure 5.5: FOC Tonnage as Share of the World Fleet, 1948-1987 (percent)

Note: FOCs include the Bahamas (1983-); Bermuda (1983-); the Cayman Islands (1985-); Cyprus (1966-); Liberia; Malta (1983-); Panama; Honduras (until 1983); Costa Rica (1953-1960 and 1975-1983); Lebanon (1960-1983); Saint Vincent (1985-); Singapore (1968-1983); and Somalia (1960-1983). The figures deviate from those in figure 5.4 because these are based upon grt and include some countries that have been excluded elsewhere, such as Singapore (1968-1983) and several minor maritime nations.

Source: Basil N. Metaxas, *Flags of Convenience: A Study of Internationalisation* (London, 1985), 17; and OECD, *Maritime Transport, 1983-1987*.

The amount of tonnage registered in FOCs temporarily peaked in 1979, although their share of the world fleet was at its highest two years earlier. One reason for the decline of FOC-registered tonnage after 1979 was increased scrapping. Some countries, such as Liberia, had a relatively large share of independent and oil company-owned tankers. Tanker tonnage comprised sixty-two percent of the Liberian fleet in mid-1980, compared with

---

[7]For an analysis of the loss records of FOC vessels, see Organisation for Economic Co-operation and Development (OECD), *Maritime Transport 1974* (Paris, 1975), 88-105.

forty-two percent for the world fleet as a whole.[8] As a result of the overcapacity of large tankers, these ships were particularly suitable for scrapping. Moreover, the decline of the Liberian fleet can be explained by increased political uncertainty. Vessels leaving the registry for this reason, however, were more likely to be transferred to other FOCs than to TMNs. Accordingly, the 20.7 million-dwt reduction in the Liberian fleet in the period 1979-1982 was partly reflected in a growth of almost nineteen million dwt in the Panamanian fleet (see figure 5.6).

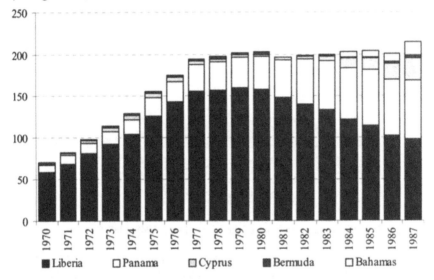

Figure 5.6: Major Flags of Convenience, 1970-1987 (million dwt)

Source: OECD, *Maritime Transport*, various years.

In 1983 Basil Metaxas claimed that the slight reduction in the importance of the FOCs was temporary and cyclical. In principle he was correct, as the reduced share of world tonnage registered in FOCs only represented a minor and temporary setback. From 1982 the share started to increase again. In addition to the old stalwarts, such as Cyprus, Liberia and Panama, new "countries" like Vanuatu, Gibraltar and the Cayman Islands entered the list.[9] In the

---

[8]OECD, *Maritime Transport 1980* (Paris, 1981), 132. The percentages are based on grt.

[9]The definition of FOCs affects the timing of their resurrection. In the 1980s some countries, such as Hong Kong, the Philippines and Singapore, were defined by the International Transport Workers' Federation as "Flags of Convenience on a Ship by Ship Basis."

latter part of the 1980s the increasing tonnage outside the traditional maritime registries was augmented by national "open registries" in many of the TMNs. Vessels registered on open registries were subject to some of the traditional maritime legislation but were granted special privileges which made them credible and cost-effective alternatives to FOCs.

The increase in the FOC fleets represented only one facet of the changing structure of the industry. Moreover, the fact that roughly two-thirds of total FOC tonnage was owned in the OECD in the late 1980s indicates that an analysis of market share overstates the reduced importance of the TMNs. Indeed, the most important feature of the transformation was not the changes in the pattern of registration of OECD tonnage but the entry into shipping of countries at a relatively early stage of economic development.

Two groups of countries increased their tonnage substantially after the shipping crisis. Although the fleets of the oil-producing nations were almost negligible at the beginning of the 1970s, they increased to approximately 3.5 percent of the world fleet by 1983. The amount of tonnage owned by these nations increased from about five million dwt in 1975 to 21.7 million dwt in 1987, peaking at roughly twenty-four million dwt in 1983.[10] Even though the relative growth was strong, these countries were fairly unimportant in an international perspective.

More significant, however, was the strong growth of non-OECD countries, of which a handful in Asia played a crucial role. The Asian share grew from less than 3.5 percent of the world fleet in 1970 to almost fifteen percent by 1987, and in tonnage terms from 11.5 million dwt in to more than ninety-three million dwt over the same period.[11] Two-thirds of the growth occurred in China, Hong Kong, the Philippines and Singapore (see figure 5.7), all of which had annual tonnage growth rates in excess of fifteen percent from 1970 to 1987.[12] The result was a massive transfer of shipping transport capacity from the TMNs to the Asian EMNs.

In many respects, the relocation of the maritime industry mirrors one of the most important development traits of the international economy since the 1960s. Just as countries that had previously been at a low level of development gained importance in the maritime sector, the momentum of world industrial

---

[10]Figures on the fleets of the oil-producing countries are from OECD, *Maritime Transport*, various years. The countries included are Algeria, Iran, Iraq, Libya, Kuwait, Saudi Arabia, the United Arab Emirates and Venezuela.

[11]The countries included in the Asian fleet are China, Hong Kong, India, Indonesia, Malaysia (1974-), the Philippines, Singapore, South Korea and Taiwan.

[12]The average annual compound growth rates were 17.5, 16.5, 15.1 and 19.9 percent for China, Hong Kong, the Philippines and Singapore, respectively.

production shifted from the OECD nations to Newly Industrialised Countries (NICs), particularly in Asia.

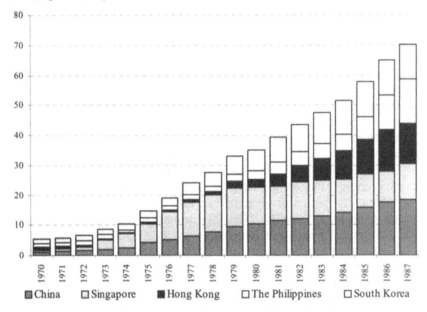

Figure 5.7: Five Most Important Asian Fleets, 1970-1987 (million dwt)

Source: See figure 5.6.

        The transformation of shipping might be explained by some of the elements used to account for changes in international manufacturing. Most analyses of the increased importance of the NICs in the international economy focus on factors such as access to a large, relatively cheap and sufficiently skilled labour force; impressive rates of investment, reflecting both high do-mestic savings and an ability to attract foreign capital and technology; sensible economic policies that facilitate saving and investments, and which sometimes targeted specific sectors; and favourable international conditions, including access to internationally mobile capital, high demand and reduced trade barri-ers. Notwithstanding that the basis of the growth differed between various countries, these elements may be relevant to addressing the changes in the maritime sector. In particular, the first three have made a positive contribution to the fleet expansion of some of the shipping nations in Asia.

        As a result of the freight market collapse, profit margins were squeezed, and shipowners were forced to focus on costs. While most costs are the same for owners of all nations, the wage bill is strongly affected by the choice of registry. In most of the TMNs domestic labour legislation made it compulsory to employ a certain number of nationals aboard vessels flying the domestic flag. The effect on costs could be substantial. In 1973 wage costs for a 6000-8000-grt dry bulk vessel in a country with low labour costs were forty percent lower than if the vessel had been flying the Norwegian flag, representing annual savings of approximately NOK 600,000.[13]

The close connection between the freight market breakdown and the increased focus on costs implies that there is a direct link between shipping crises and changes in the international shipping hegemony. When shipping demand was high and rising, revenues were sufficient to allow high-cost countries to offer competitive services despite higher variable costs. Moreover, high investment levels in TMNs enabled the build-up of technological superiority. When the freight market collapsed and investments dried up it became increasingly difficult for high-cost nations to compete and operate profitably at the prevailing level of freight rates.

In several Asian countries the existence of a relatively cheap labour force with maritime training and experience stimulated the growth of domestic fleets. Some countries had long traditions of providing seamen for vessels flying foreign flags. Personnel from South Korea, the Philippines and Taiwan comprised more than forty-six percent of the crews on Liberia-registered vessels in 1978.[14] In addition to bringing valuable foreign exchange to the domestic economy, some seamen used their income and experience to establish their own maritime businesses. The access to skilled, low-cost labour has consequently been significant for the growth of some of the most important EMNs.

High rates of saving and investment were also important in the growth of the NICs, both in relation to other developing countries and to the industrialised nations. Capital was acquired through a combination of high domestic savings, debt accrued abroad and foreign direct investment. The latter was particularly important in the maritime sector. Several shipping companies shifted some or all of their operations from Europe and North America to Asia, partly in response to the booming regional trade and partly due to cost factors. The capital and skills brought to the area by these firms were important in the development of domestic shipping, and the presence of foreign subsidiaries supplemented its growth.

---

[13]Norway, Parliament, Stortingsmelding No. 23 (1975-1976), "Om sjøfolkenes forhold og skipsfartens plass i samfunnet" ["On the Conditions for Seamen and Shipping's Role in the Society"], 62.

[14]Figures from Norges Offentlige Utredninger (1983:7), "Skipsfartens konkurranseevne" ["The Competitiveness of Shipping"], 54.

Measuring foreign direct investment in shipping is difficult.[15] The fact that foreign companies established sales offices, subsidiaries or shipowning companies in the EMNs was crucial for the growth of some of them as maritime centres. Moreover, the development of some Asian fleets was the direct result of strong links to other nations, as in the case of the *shikumisen* deals between Hong Kong shipowners and Japanese corporations. Domestic investments in shipping, and the integration of Asia into the international ship financing market, enhanced the growth of locally-based shipping.

A common element in the growth of the EMNs was the attempt to encourage the entry of foreign maritime firms and the expansion of local companies through specific policies. In the manufacturing sector, the extent and effects of such policies are controversial. In shipping, however, it is fairly easy to see that such policies have been successful in expanding the fleets of certain countries.

Singapore is perhaps the best example of this relationship. In 1968 the Singapore government decided to open its flag to shipowners of all nations. This was modelled on measures similar to those in traditional FOCs. But in Singapore the authorities tried to establish a relatively close link between the national economy and shipping, especially through tax rebates to shipowners employing domestic crews. The success of the scheme was overwhelming: the Singapore fleet increased from 230,000 grt to almost eight million grt in a ten-year period. In an effort to clean up the image of the Singapore flag, in the late 1970s the government introduced stricter regulation of the country's fleet. The shipping policy had nevertheless been successful in creating a strong national fleet and a viable shipping environment, augmented by the importance of Singapore as a regional maritime hub. Moreover, the introduction of higher standards did not lead to any significant flight of tonnage. In 1987 Singapore ranked thirteenth among the world's maritime nations – ahead of the United Kingdom, its previous colonial master.

The provision of labour and capital, and the effect of specific policies, seems to have been important for the growth of Asian manufacturing and shipping. The final elements commonly used to explain the growth of manufacturing in the NICs – access to mobile international capital, high demand and reduced trade barriers – applied only to a limited extent to shipping. Although access to international capital was important, it would be foolhardy to claim that the shipping sector was subject to high demand. Indeed, the opposite was the case: demand for shipping services was falling. Nevertheless, the depressed state of the shipping market may be important in explaining the shift of the world fleet from the TMNs to the EMNs and FOCs.

---

[15]See Stig Tenold, *Norwegian Shipowning Companies and Foreign Direct Investment* (Bergen, 2000), for a suggestion on how foreign direct investment in shipping might be analysed.

The combination of the increased focus on costs and the reduced competitiveness of the TMNs was perhaps the single most important factor in explaining the transfer of vessels to FOCs and EMNs. The shipping policies of several of the most important maritime nations were changed following the freight market collapse. The use of low-cost foreign labour was facilitated by increased access to registry abroad and a relaxation of domestic manning requirements.[16] Judging by the shifts in the world fleet, the former option was particularly attractive. In the late 1980s, US companies owned tonnage registered under thirty-six flags, and the UK, Japan, Germany, Norway, Greece, Hong Kong and the Netherlands also operated tonnage on more than twenty different registries.

Reduced trade barriers have been used to explain the growth of manufacturing in the NICs. The experience in shipping is somewhat different. On the one hand, the EMNs may have benefited from the relaxed barriers to trade in offering bulk shipping services to and from OECD countries, as well as the indirect effect of trade liberalisation on the demand for seaborne transport. At the same time, however, several countries expanded their own shipping activity through restrictions on the transport of domestic imports and exports. At the beginning of the 1980s, the United Nations Conference on Trade and Development's (UNCTAD) Code of Conduct, with its forty-forty-twenty cargo-sharing arrangement, was adopted after fifty countries had ratified the agreement.[17] This arrangement enhanced the possibilities of EMNs developing domestic liner fleets.

In addition to cargo reservation, some EMNs used subsidies to build up their domestic merchant marines. Even though the overall costs of such schemes generally exceeded the benefits, cargo reservation and fleet subsidies have been successful in increasing the size of domestic fleets. The cargo-sharing arrangements and the use of subsidies do not correspond well with the idea of reduced barriers to trade. Nevertheless, their effect may be seen as positive when the expansion of the domestic fleets – rather than the net cost to the countries – is used as a criterion.[18] Moreover, there have been barriers to

---

[16]In the case of the Norwegian International Ship Register (NIS), established in 1987, the only condition is that the head officer be a Norwegian national, although this requirement may be waived.

[17]The number of countries necessary to implement the Code, set at twenty-four, had been achieved by early 1978. But the additional requirement that the contracting parties should own at least twenty-five percent of world tonnage was not met until the European Economic Community ratified it.

[18]See Alexander J. Yeats, *Shipping and Development Policy: An Integrated Assessment* (New York, 1981), for an introduction to some of the economic aspects.

trade in manufactured goods that are similar to the protective mechanisms util-ised in shipping.

Another reason for the increased attractiveness of EMNs and FOCs may have been the reduction of vessel standards. As a result of overcapacity, investments in new tonnage slowed since they were unlikely to be profitable, and it became increasingly difficult for shipping companies to accumulate the funds for fleet modernisation and expansion. Yet shipowners were unwilling to scrap existing tonnage in the hope that a market upturn would generate increas-ing demand for both transport services and second-hand tonnage.

The declining construction of new vessels and the reluctance to scrap older ones led to an increase in the average age of the world fleet and an ap-parent increase in casualties. Tonnage older than ten years increased from forty-one percent in 1980 to sixty-one percent in 1987, while tonnage lost grew from about one million grt (0.25 percent of the world fleet) in 1977 to 2.7 mil-lion grt (0.6 percent) in 1986.[19] The casualty figures, however, fluctuated sharply, and Hans Peters claims that the loss ratio doubled between 1988 and 1991.[20] Although there is no clear trend in the share of the world fleet lost, it is evident that the FOCs were responsible for a disproportionate share.

The reduction in vessel standards may have prompted transfers from TMNs to FOCs, as some of the latter were often regarded as having more le-nient attitudes toward seaworthiness. This claim is difficult to substantiate, but it may have played a minor role in explaining the shift in the maritime hierar-chy. Yet the main reasons for this shift were the factors described above, with a particular emphasis on the need to reduce costs in a depressed market.

Helen Thanopoulou has offered an alternative explanation of the shift in the international distribution of the shipping industry.[21] She has suggested that there is a feedback relationship between the entrance of less-developed countries into shipping and the prolongation and deepening of maritime crises. In many respects her analysis is similar to the one presented above, but her focus is on the dynamics of the international economy rather than specific fea-tures of the EMNs. The basis for her analysis is that shipping depressions ad-versely affected the ability of the higher-cost nations to remain in the market. TMNs attempted to increase specialisation, focusing on capital-intensive seg-

---

[19]Figures for age distribution are from table XVIII and for losses from table XX of OECD, *Maritime Transport* (Paris, 1981 and 1987).

[20]Hans J. Peters, *The Maritime Transport Crisis* (Washington, DC, 1993), 23. The loss ratios cited are, however, far smaller than those calculated based on figures from the OECD, as used in my example.

[21]Helen A. Thanopoulou, "The Growth of Fleets Registered in the Newly-emerging Maritime Countries and Maritime Crises," *Maritime Policy and Manage-ment*, XXII, No. 1 (1995), 51-62.

ments to maintain a competitive advantage. When this proved unsuccessful, the EMNs made a dynamic entry into all shipping segments, regardless of capital intensity. According to Thanopoulou, this challenge can be explained by the fact that shipping services had become relatively standardised, following the Vernon product cycle.

Gunnar Sletmo has also used the Vernon product cycle to examine the shipping industry. At the introductory stage, the product – shipping services – was not standardised, implying that it was affected by technical and commercial experimentation. As the product matured it became more standardised, a fact that changed the competitive advantage of various suppliers. According to Sletmo, shifts in the distribution of the world fleet are the result of the maturity of the bulk sector: "the impossibility of long-term product differentiation in bulk shipping and the high degree of substitutability of highly mobile services make the search for least-cost production systems even more essential in shipping than in manufacturing."[22]

The maturation of shipping services within a Vernon product cycle framework provides one explanation for the alterations in the hierarchy of international shipping. In many ways, the services offered by the industry evolved in a manner that rewarded the properties and factor relationships found in several Asian countries. For instance, the sharp fall in vessel values greatly lowered the investments needed to participate in shipping and in essence reduced the capital intensity of several shipping segments. Simultaneously, vessel ownership in TMNs became unfavourable, following developments in the shipping market, such as lower freight rates, and in the domestic economy, such as increasing wages for seamen.

It is thus evident that the decline of the TMNs can be understood in the context of long-term trends that are familiar from other sectors. Moreover, factors in both the countries of reduced importance and those that increased their share of the world fleet were essential. Changes in the international division of labour, the maturation and standardisation of products and services and the increasing importance of NICs combined with the reduced viability of OECD shipping to impart large structural changes to the maritime sector.

The transformation did not affect all TMNs equally, and in some countries the shipping crisis had particularly adverse effects. We may expect a more substantial reduction of the fleets that were severely hit by the crisis, as unprofitable tonnage was sold abroad and vessels owned by companies in economic distress were taken over by foreign creditors. Moreover, we might expect that shipowners who were less severely affected would consolidate their position or increase their activities. This would be reflected in a stable or growing share of world tonnage. Figure 5.8, which shows the development of the fleets of the five most important maritime nations in 1970, excluding the

---

[22]Sletmo, "Shipping's Fourth Wave," 296.

US, illustrates both the differences in the development of the various countries and the influence of specific circumstances. The strong decline of the Norwegian and British fleets reflects the problems of domestic shipping companies, which resulted in large sales of tonnage and the reduced attractiveness of the domestic flag. The growth of the Greek fleet up to the beginning of the 1980s was in part due to the transfer of Greek-owned vessels from foreign registries. From 1974 to 1981 the development of the Greek-owned fleet under Greek and other flags went in opposite directions.[23] Similarly, the reduced importance of the Liberian fleet, particularly from 1984, was a result of political uncertainty and was more than compensated for by the increase of other FOCs.

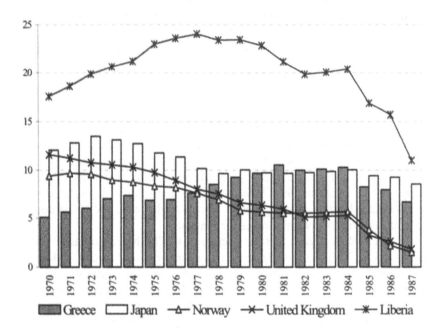

Figure 5.8: Development of the Five Most Important Fleets, 1970-1987 (percent of world dwt)

Source: See figure 5.6.

The development of the Norwegian and British fleets stands out when compared with the other main maritime nations. In 1970 the UK and Norway had the world's third and fourth largest fleets, respectively. The Greeks caught up with the Norwegians in 1976 and passed the UK the following year. By

---

[23]See Gelina Harlaftis, *A History of Greek-owned Shipping: The Making of an International Tramp Fleet, 1830 to the Present Day* (London, 1996), 266.

1981 Norway and the UK had been relegated to fifth and sixth place due to the growth of the Panamanian fleet. By 1987 the UK ranked as the fourteenth largest maritime power, with Norway eighteenth. The British had been overtaken by the US and USSR; Asian EMNs such as China, the Philippines, Hong Kong and Singapore; the FOCs, such as Cyprus and Bahamas; and even by Italy. Norway also trailed South Korea, Brazil and India. Measured in tonnage, the flight from the Norwegian and British flags was greater than for any other domestic flag. Both fleets peaked in late 1976, and by the middle of 1987 the Norwegian and British fleets had been reduced by approximately forty million dwt compared with the apex.[24] Due to differences in the size of the fleets of various countries, absolute figures do not necessarily allow a meaningful comparison of their development. Still, the large reductions of Norwegian and British tonnage are distinct even when relative figures are used.

Table 5.1
The Development of the Largest OECD Fleets, 1973-1987

Country	Fleet 1973 ('000 dwt)	Fleet 1980 ('000 dwt)	Fleet 1987 ('000 dwt)	Change 1973-1980	Change 1980-1987	Change 1973-1987
Denmark	6512	8703	6961	33.6	-20.0	6.9
France	13,286	20,861	8407	57.0	-59.7	36.7
Germany	12,148	13,332	5659	9.7	-57.6	53.4
Greece	31,438	67,048	42,776	113.3	-36.2	36.1
Italy	13,185	17,951	12,178	36.2	-32.2	-7.6
Japan	58,585	67,321	54,669	14.9	-18.8	-6.7
Netherlands	7264	8999	5123	23.9	-43.1	29.5
Norway	40,087	38,885	9657	-3.0	-75.2	75.9
Sweden	8802	13,522	2403	53.6	-82.2	72.7
UK	47,155	43,814	11,677	-7.1	-73.3	75.2

Note: Includes the ten largest OECD fleets in 1970, representing all vessels greater than 100 grt. The US fleet is excluded.

Source: See figure 5.6.

Table 5.1 illustrates the reduction of the OECD fleets and the severity of the contraction in Norway and the UK. One noteworthy aspect is that the Norwegian and British fleets were smaller than their 1973 levels as early as 1980. The fleets of the other main OECD maritime nations increased, considerably in several countries, between 1973 and 1980. There might be two reasons for this disparate development. One may be that shipowners in Norway and the UK were particularly hard hit by the freight market collapse and that

---

[24]Fearnley and Eger, *Review*, uses a narrower definition of fleet size; when this definition is followed, the Norwegian fleet peaked at the beginning of 1977.

liquidity problems and bankruptcies therefore occurred earlier, expediting the sale of domestic tonnage to foreigners. Alternatively, it may be that shipowners in these countries realised the severity of the crisis earlier than their competitors and reacted by disposing of tonnage which would not be profitable under the domestic flag.

A closer examination of Norwegian sales indicates that with regard to the Norwegian shipping industry the former explanation is more relevant. In the British case, approximately two-thirds of tanker tonnage was owned by oil companies, and the reduction of the British fleet was enhanced by the transfer of oil company tonnage from the British flag to FOCs. Moreover, a considerable share of the fleet flying the Red Ensign in the early 1970s was owned by foreigners who wanted to benefit from the UK's liberal investment grants. When the conditions for these grants had been fulfilled after five years, the owners chose to transfer the vessels to foreign flags.[25]

The relative reduction of the British, Norwegian and Swedish fleets stands out in long-term perspective. In all three countries liberalised flag policies led to a reduction in the domestically-registered fleet, which means that the decline was less than table 5.1 suggests. Both relative and absolute figures show that the decrease in the domestic fleet was more severe in Norway, which suggests that its shipping sector was particularly hard hit by the crisis.

One approach to judging whether Norwegian shipowners were particularly hurt by the crisis would be to compare the revenues and profits of Norwegian shipowning companies with those of foreign firms. Although such an analysis would enable us to judge how the economic performance of Norwegian shipping companies rated internationally, it is an extremely difficult task. Precise accounting details are difficult to find and are seldom comparable across national borders. The same applies to figures for bankruptcies and liquidations, as the institutional approach to firms in distress varied.[26] An analysis of the entry and exit of shipowning firms is complicated by structural differences: in some countries it was common to establish a company for each ship, while in others all tonnage was registered as owned by a single firm. Moreover, in several instances shipping companies were divisions of businesses operating primarily in other sectors. Any study of shipping revenues and structural developments would consequently be fraught with uncertainties, and the results would largely be determined by the definitions and methods used.

The absence of a suitable basis for international comparison makes it hard to find a good method to evaluate the fate of Norwegian shipowners in an international perspective. The above-average declines for Norway, the UK and

---

[25]Ronald Hope, *A New History of British Shipping* (London, 1990), 450.

[26]Due to agreements utilised in the shipping industry, such as time charters, a bankruptcy will be relatively unfavourable for the main creditors, who thus have an incentive to find other solutions.

Sweden shows that all were adversely affected. Still, the tonnage decline was influenced by long-term trends outside shipping.[27] Although fleet growth or decline can have a number of causes, we can augment this indicator with other factors, such as lay-up rates, in order to gauge the magnitude of the crisis.

Norwegian lay-ups during the shipping crisis were higher than for any other major fleet. This dubious record indicates the differences in economic performance between Norwegians and their rivals. Shipowners in all countries were faced with falling revenues due to decreasing freight rates. Laid-up vessels, however, were unable to earn any revenues unless they were laid-up on the charterer's account. The result of the high lay-up rates was that Norwegian shipowners, to a larger extent than foreigners, were affected by the loss of revenue caused by the lack of employment for their vessels.

Figure 5.9 is based on estimates from the General Council of British Shipping, the Norwegian newspaper *Norges Handels og Sjøfartstidende* and Fearnley and Eger.[28] The exact share of tonnage laid-up is difficult to estimate, particularly when the shifts were large. In August 1975, the General Council of British Shipping reported that 29.6 million dwt of tanker and combination carrier tonnage was laid-up, whereas the British shipbroker John I. Jacobs estimated that it was approximately forty-four million dwt.[29] Moreover, the lay-up rate increases if we look at dwt, since the majority of such tonnage comprised larger vessels. Although the various figures deviate, the differences are generally fairly small, and there is no reason to doubt that the Norwegian lay-up rate from 1975 onwards was considerably above international norms.

In late 1974 TT *Sir Charles Hambro* and TT *Julian* were the first mammoth tankers officially laid-up.[30] As the crisis evolved, lay-ups affected a substantial share of Norwegian shipowners. Figure 5.9 illustrates the significant differences between Norwegian and international lay-up rates. Indeed, the contrast is so large that the effects on Norwegian shipping companies must have been far more severe than for foreign firms. By the summer of 1975 one-quarter of the laid-up tankers and combination carriers flew the Norwegian flag. Figures from the British shipbroker E.A. Gibson indicate that whereas

---

[27]One such factor would be the appearance of more profitable investment alternatives, making a reduction of the fleet rational. Changes in public policies may be another reason for differences in the various countries.

[28]These sources generally correspond fairly well. In the quarters where the figures deviate, an estimate based on all three sources has been constructed.

[29]Norges Rederforbund, *Momenter til belysning av norsk og internasjonal skipsfart* [*Aspects Illuminating Norwegian and International Shipping*] (Oslo, 1975), 18.

[30]H.P. Drewry Shipping Consultants Ltd., *The Trading Outlook for Very Large Tankers* (London, 1975), 53.

the international lay-up rate for tankers and combination carriers was ten percent, the corresponding Norwegian figure was twenty-five percent.[31]

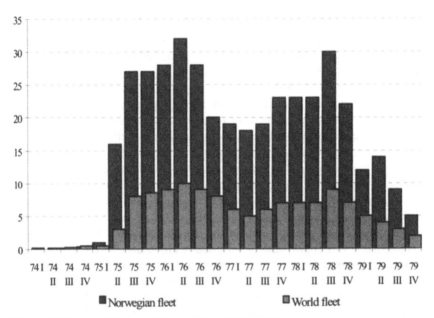

Figure 5.9: Lay-up Rates as a Share of Dwt, 1974-1979

Note: Figures refer to the beginning of each quarter.

Source: See text.

By December 1975, forty-six Norwegian shipowners owned laid-up tankers amounting to more than ten million dwt, or over forty percent of the tanker fleet.[32] Of the eighty-two laid-up oil tankers, only eight (1.3 million dwt) were laid-up on the charterer's account, and only four owners of such vessels had not been forced to lay-up ships on their own account as well. The average size of laid-up Norwegian tankers was about 130,000 dwt, and twenty-four were larger than 200,000 dwt. These were a combination of older vessels, such as the relatively small sixteen-year-old *Bente Brøvig*, and modern mammoths, such as the 311,400-dwt *Belfri*, delivered from Germany in March

---

[31]Figures from Gibson quoted in *Norges Handels og Sjøfartstidende*, 1 July 1975. The data refer to vessels laid-up on the owner's account.

[32]See the list in appendix 1 for a detailed account of laid-up vessels, owners and the date they were idled.

1975 and immediately laid-up. When Norwegian lay-ups peaked in the spring of 1976, 116 vessels, amounting to more than fourteen million dwt and representing almost one-third of Norwegian tonnage, were mothballed.

Table 5.2

Fleet Size and Lay-up Rates, 1974-1979 (grt)

Country	Fleet 1974 ('000)	Fleet 1978 ('000)	Lay-up 1974	Lay-up 1975	Lay-up 1976	Lay-up 1977	Lay-up 1978	Lay-up 1979	Average Lay-up
Denmark	4505	5530	0.0	0.8	11.2	8.1	10.0	3.0	5.5
France	8835	12,197	0.0	0.9	1.9	0.8	2.1	1.2	1.2
Germany	7980	9737	0.0	0.0	4.1	4.6	6.8	3.0	3.1
Greece	21,759	33,956	1.0	1.2	5.9	7.4	9.3	2.6	4.6
Italy	9322	11,492	0.7	0.6	9.6	7.2	4.6	1.8	4.1
Japan	38,708	39,182	0.0	0.0	0.9	0.1	1.5	7.0	1.6
Netherlands	5501	5180	0.0	0.4	8.1	9.0	0.0	0.0	2.9
Norway	24,853	26,128	0.0	1.0	24.3	17.1	21.4	6.1	11.7
Spain	4949	8056	0.4	0.4	0.3	0.1	0.7	0.6	0.4
Sweden	6227	6508	0.0	1.6	27.3	8.3	25.3	6.3	11.5
UK	31,566	30,897	0.0	0.1	8.4	4.8	4.9	3.2	3.6
Liberia	55,322	80,191	0.3	0.5	10.6	7.3	7.6	2.3	4.7
Panama	11,003	20,749	0.7	0.0	3.8	1.4	1.9	0.8	1.4

Note: Includes all countries with fleets of more than five million grt in 1975, except for the US, which is omitted due to problems in calculating the reserve fleet, and India, Singapore and the USSR, for which lay-up data are unavailable. The fleet at the beginning of each year is measured as the average of the fleets in the middle of the reported year and the middle of the previous year in '000 grt. The difference in the Norwegian lay-up figures between this table and figure 5.9 is due to the fact that the latter is based on dwt and this table on grt.

Source: OECD, *Maritime Transport*, various years.

A total of seventy-three Norwegian companies owned oil tankers in 1976, and more than sixty percent had been forced to lay-up some or all of their vessels. This illustrates the breadth of the crisis among Norwegian owners. The shipowners affected were a combination of small owners, such as T.S. Bendixen AS, with all their tonnage laid-up; major owners, such as Hilmar Reksten, with large parts of their fleets idled; and important owners, such as Sig. Bergesen d.y. and Wilh. Wilhemsen, with small shares of their fleets laid-up. The most affected were traditional shipping companies that had played an important role in the postwar expansion of Norwegian shipping.

An international comparison of lay-up rates justifies the assertion that Norwegian shipowners were hit harder by the crisis than their rivals elsewhere (see table 5.2).[33] As a share of dwt, Swedish lay-ups were at times even higher, but in terms of tonnage this amounted to only a third of Norwegian lay-ups, since the Norwegian fleet was much larger than the Swedish. Liberia was the only nation that laid-up as many dwt as Norway, but since its fleet was bigger, the actual lay-up rate was smaller.

Lay up-figures indicate the huge problems encountered by Norwegian shipowners when attempting to secure profitable transport assignments. Table 5.2 shows that, with the exception of Sweden, the average proportion of the fleet laid-up by Norwegian owners was more than twice as high as in other important maritime nations. The reasons for this will be analysed in detail later. But here it is important to note that domestic conditions affected the lay-up figures of some countries. The principal example is Japan, where vessel lay-ups were impeded by seafarer employment guarantees which made the practice less beneficial than continued operations.[34]

---

[33]In 1976, the oil-producing nations were hit most severely. Iraq, Saudi Arabia and Libya had been forced to lay up twenty-six, thirty-nine and seventy-four percent of their tanker fleets, respectively. But these were rather small: in mid-1976 together they amounted to less than ten percent of Norwegian capacity; see OECD, *Maritime Transport 1975* (Paris, 1976), 77; and *Maritime Transport 1976* (Paris, 1977), 123.

[34]OECD, *Maritime Transport 1975*, 101.

Chapter 6
The Basis for the Crisis in Norwegian Shipping

International shipping is by definition a very mobile industry. The primary means of production are vessels traversing national borders, and owners do not have to conform to the same restrictions on location as do agents in many other markets. Shipping has gradually become detached from national markets with regard to the provision of capital and labour and consequently has acquired an increasing share of its factors of production in international markets. Nevertheless, a large proportion of transport capacity is owned in a handful of important shipping nations. Norwegian shipowners have played an important role in the international shipping market since the mid-nineteenth century.

The importance of Norway as a major maritime nation was also reflected in support functions such as shipbroking, insurance, research and classification. Moreover, there has been a relatively strong relationship between Norwegian shipowners and the domestic shipbuilding and ship equipment industries. Despite the fact that shipowners purchase their vessels in the international market, Norwegian shipping has functioned as a catalyst for, and been an important purchaser from, the domestic shipbuilding industry.

Norway's significant international role was mirrored in the importance of shipping to the Norwegian economy. In 1974 domestic/coastal and foreign-going vessels accounted for nine percent of the country's capital stock and four percent of employment. The foreign-going fleet is more capital intensive than the coastal and uses non-Norwegian seamen to a greater extent, so the difference between capital and employment would be even more marked if only the deep-sea merchant marine is considered. The fleet operating in international waters accounted for 2.6 percent of Norwegian employment.

The shipping sector has always been important to the Norwegian economy.[1] Yet during the last 200 years it has changed from mainly transporting Norwegian imports and exports to a service industry in which most of the

---

[1]Fritz Hodne, *Norges økonomiske historie 1815-1970* [*The Economic History of Norway, 1815-1970*] (Oslo, 1981), 132, presents shipping in the period 1815-1914 as "the third large export sector." The two other were fish and timber. Odd Aukrust, *Norges økonomi etter krigen* [*The Norwegian Economy after the War*] (Oslo, 1965), 190, claims that "[t]here may be reason to emphasise the great extent to which the [foreign trade] figures in the postwar period have been influenced by the development in the shipping sector."

demand originates abroad.[2] As shipping services have mainly been sold internationally, it has been crucial to the Norwegian balances of trade and payments.

Gross freight earnings – the remuneration from the sale of transport
services – accounted for somewhere between thirty-five and fifty percent of
total Norwegian exports in most years from 1870 to 1970.[3] Put another way;
foreign revenues from Norwegian shipping facilitated between a third and a
half of imports and largely neutralised the merchandise trade deficit. Naturally,
a considerable part of the income was spent by shipowners in purchasing vessels from foreign yards or on costs accrued abroad.

The liner sector has accounted for the most stable share of Norwegian
shipping revenue (see figure 6.1). Despite the fact that the share of the fleet in
the liner trades fell as low as five percent in the 1970s, its share of gross
freight earnings accounted for more than twenty percent of aggregate Norwegian earnings in each year of the decade.[4] But this large share of gross earnings was offset in part because the ships operated in a high-cost environment,
and their relative importance is reduced if costs are considered in an analysis
of freight earnings. This will be the most fruitful approach in a discussion of
profitability.

Gross freight earnings also reveal the reduced importance of the charter market after the oil price increase. The difficulties in securing profitable
charters led to a fifty percent reduction in gross charter earnings between 1974
and 1979. Moreover, there was an additional deterioration of profits due to the
increased costs of bunkers and crew. According to current account figures,
expenditures abroad by Norwegian ships almost doubled between 1973 and
1979, whereas gross freight earnings increased by only about a third.[5]

In the 1970s, fluctuations in gross freight earnings for tankers were
particularly large since the shipping crisis led to lower freight rates and reduced employment. Although the Norwegian tanker fleet, measured in deadweight tons (dwt), increased by forty-six percent from 1973 to 1977, tanker

---

[2]Camilla Brautaset, "Norsk eksport 1830-1865 i perspektiv av historiske nas-
jonalregnskaper" ["Norwegian Exports 1830-1865 in an Historical National Accounts
Perspective"] (Unpublished Dr. oecon. thesis, Norwegian School of Economics and
Business Administration, 2002).

[3]In most of these years gross freight earnings contributed in excess of forty
percent of total exports.

[4]Norway, Statistisk Sentralbyrå, *Historisk Statistikk 1994* [*Historical Statistics
1994*] (Oslo, 1994), table 20.18.

[5]*Ibid.*, table 22.8. It is important to remember that an abrupt halt in chartering
will only be felt gradually since it takes some time before previous charters expire.

earnings were reduced by a quarter. Overall, however, tankers were responsible for the largest share of Norwegian gross freight earnings in the period 1960-1980 (see figure 6.2).

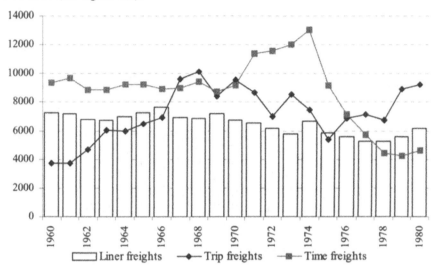

Figure 6.1: Gross Freight Earnings by Norwegian Vessels, 1960-1980 (million 1980 NOK)

Note: These figures are deflated by the Norwegian consumer price index, using 1980 as the base year. They differ from the gross freight earnings presented in Norway, Parliament, "Utenriksregnskap: Driftsregnskap overfor utlandet" ["Balance of Payments: The Current Account"], table 22.8. The figures used here are based on replies from shipowners on gross freight earnings for ships of 500 gross registered tons (grt) and over employed in international trade.

Source: Norway, Statistisk Sentralbyrå, *Historisk Statistikk 1994 [Historical Statistics 1994]* (Oslo, 1994), table 20.18.

The "trade balance" of the shipping sector fell considerably after 1974.[6] Shipping's contribution to Gross Domestic Product was reduced from 8.6 percent in 1973 to 3.8 percent in 1977.[7] The contribution to Net Domestic Product was even smaller as a result of the large depreciation of the capital

---

[6]Arnljot Strømme Svendsen, "Skipsfartskonjunkturene i 1970-årene" ["Shipping Cycles in the 1970s"], *Sjøfartshistorisk Årbok 1978 [Norwegian Yearbook of Maritime History 1978]* (Bergen, 1979), 240.

[7]Norway, Statistisk Sentralbyrå, *Historisk Statistikk 1994*, table 22.12.

equipment utilised in the shipping sector, falling from 5.9 percent in 1973 to 1.1 percent four years later.[8]

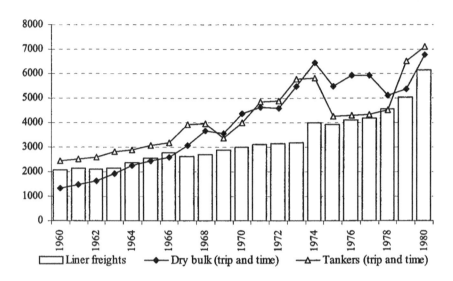

Figure 6.2: Gross Freight Earnings by Vessel Type, 1960-1980 (million NOK)

Source: Statistisk Sentralbyrå, *Historisk Statistikk 1994*, table 22.8.

The increasing export surplus after 1977 was mainly a result of stable income from the sales of second-hand vessels coupled with a reduction in ship imports (see table 6.1). In 1978 the export of second-hand ships was higher than Norwegian vessel purchases abroad, which suggests that the trade in vessels contributed positively to the sector's surplus and that a smaller share of the surplus was due to actual shipping services. Although this trend shifted in the late 1970s, in the first half of the 1980s the pattern was clear: tonnage sales, reflecting the reduced activity in Norwegian shipping, comprised a larger part of the export surplus, and the provision of shipping services became less important. Shipping lost its position as the most important export sector to the oil industry in the late 1970s.[9]

---

[8]The sector had less than ten percent of the Norwegian capital but accounted for twenty-five percent of the consumption of fixed capital.

[9]Cf. Norway, Statistisk Sentralbyrå, *Historisk Statistikk 1994*, table 22.8.

Table 6.1
Export Surplus of Shipping, 1970-1981 (million NOK)

	1970	1971	1972	1973	1974	1975
Gross receipts	12,490	13,990	14,175	16,705	19,220	16,720
Expenditure abroad	4700	5695	5540	6450	7935	7970
Ship imports	2978	3771	3478	5884	4959	6586
Sales of ships (second-hand)	1369	802	2242	3798	4072	2946
Surplus	6181	5326	7399	8169	10,398	5110
Surplus (1980 NOK)	13,833	11,223	14,451	14,934	17,393	7634

	1976	1977	1979	1980	1981	1981
Gross receipts	17,325	17,470	18,205	22,255	26,980	30,934
Expenditure abroad	8895	9495	9960	12,769	16,265	18,894
Ship imports	8103	7639	3352	3443	1434	4541
Sales of ships (second-hand)	2579	3440	3536	3417	2425	3406
Surplus	2906	3776	8429	9460	11,705	10,905
Surplus (1980 NOK)	3979	4740	9803	10,494	11,705	9590

Source: Norway, Statistisk Sentralbyrå, *Historisk Statistikk 1994*, tables 18.2, 18.3 and 22.8.

Norway's position in the international shipping industry was in free fall during the shipping crisis. Indeed, more than three-quarters of the Norwegian fleet was transferred to foreign companies. Moreover, shipping's dominant position in the Norwegian economy was seriously challenged. Why did Norwegian shipping fail so spectacularly?

The answer to this question can be found in the strategies pursued by Norwegian shipowners around the time of the freight market collapse. The use of aggregate figures in this analysis, however, conceals the heterogeneity of Norwegian shipping. Yet as one industry insider pointed out in 1979, "there is no shipowning company of any importance that has not been influenced, and in most cases negatively so, by the five year long crisis in shipping."[10] The dominant Norwegian strategy – to focus on large tankers and bulk carriers, spot-market chartering and aggressive contracting – served the industry well in a market in which demand increased rapidly, but it left the shipowners vulnerable when conditions shifted. The high rate of Norwegian lay-ups and the sharp reduction of the fleet in the wake of the freight market breakdown indicate that Norwegian shipowners were severely affected by the crisis.

This misfortune can be explained by the owners' business strategies. The term embodies a plethora of decisions. For example, the decision to order

---

[10]Leif Terje Løddesøl, "Hvorfor gjør noen rederier det godt og andre det dårlig?" ["Why Are Some Shipping Companies Successful, and Some Not?"], *Internasjonal Politikk*, No. 1B (1979), 167.

a vessel is an obvious example of a strategic judgement. Even though it is often made less consciously, the decision not to dispose of existing tonnage is another example. The variety and volume of evaluations and decisions that comprise any strategy make it difficult to present a uniform and all-encompassing portrait. Instead, this analysis will focus on three elements – fleet structure, chartering policy and contracting – which together embody the most important strategic decisions in shipping. The aim of the analysis is twofold. The first is to identify the factors that characterised these three elements, that is, to find out what was unique about Norwegian fleet structure, chartering policy and contracting through international comparison. The second object is to analyse the underlying factors and forces that shaped the strategic decisions.

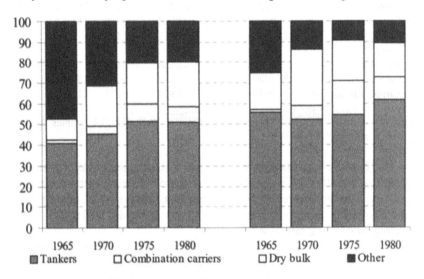

Figure 6.3:   Fleet Composition as a Percentage of Dwt, 1965-1980 (World [left] and Norway [right])

Source: Fearnley and Egers Chartering Co. Ltd., *Review*, various years, tables 3 and 20.

Fleet structure includes vessel type, size, propulsion and age. One reason for the high rate of Norwegian lay-ups was its disproportionate number of vessel types and sizes affected adversely by the market imbalance. The largest mismatch was in tankers, where Norwegian shipowners held a disproportionate share of the world's tonnage.[11] The effects of the freight market collapse were least visible among liners, which comprised a small proportion of

---

[11]A disproportionate share implies that Norwegian ownership of these vessels was larger than its share of the total world fleet.

the fleet. Figure 6.3 shows the distribution of the international and Norwegian fleets across vessel types. Norway's high investments in the bulk sector are evident, and this distinctive structure was apparent as early as the mid-1960s. Table 6.2 depicts the considerable differences in the composition of various national fleets. Liberia was the only major maritime nation with a larger share of tankers, combination carriers and dry bulk vessels than Norway.

Table 6.2
Fleet Composition, Seven Largest Fleets, 1976 ('000 grt)

Flag	Oil Tankers	Dry Bulk/ Combination	Sum Bulk	Misc.	Fleet 1976
Liberia	64	29	93	7	73,447
Japan	46	31	77	23	41,663
UK	49	25	74	26	32,923
Norway	53	34	87	13	27,944
Greece	36	33	69	31	25,035
USSR	20	4	24	76	20,668
Panama	38	21	59	41	15,631

Source: Organisation for Economic Co-operation and Development (OECD), *Maritime Transport 1976* (Paris, 1977), 124.

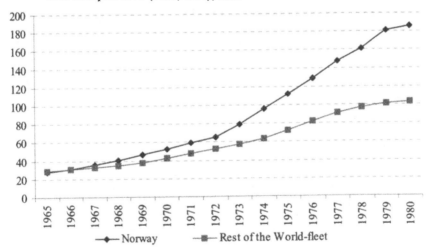

Figure 6.4: Average Tanker Size, 1965-1980 ('000 dwt)

Note: Figures for the Rest of the World have been calculated by subtracting Norwegian tonnage from the figures for the world fleet. If Norwegian tonnage was included, the figures for the world fleet after 1967 would be higher (but not significantly).

Source: Fearnley and Egers Chartering Co. Ltd., *Review 1972* (Oslo, 1973); and *Review 1980* (Oslo, 1981), tables 3 and 20.

Another feature of the Norwegian fleet, partly a result of the focus on bulk vessels, was the high proportion of large vessels (see figure 6.4). Norwegian owners had invested heavily in the size classes that operated in the hardest-hit market segment. Mammoth tankers, combination carriers and bulk carriers comprised a larger share of the Norwegian than the international fleet. Figure 6.4 is based on tankers of more than 10,000 dwt and shows the large and increasing difference between the average size of Norwegian tankers and the rest of the international tanker fleet.[12] Whereas Norwegian tankers were twenty-two percent larger than non-Norwegian ships in 1970, the corresponding figures for 1973 and 1980 were thirty-eight and more than eighty percent, respectively. Due to the larger share of tankers in the Norwegian merchant marine, average size differences were even greater for the total fleet.

Table 6.3 shows that vessels above 50.000 grt comprised around fifty-five percent of the Norwegian fleet in 1975 compared with roughly a third for the world fleet. Four years later more than two-thirds of the Norwegian fleet consisted of vessels larger than 50,000 grt compared with thirty-eight percent for the world fleet. The average size of Norwegian vessels was more than twice that in the world fleet in both instances.

Table 6.3
Size Distribution, Norwegian and World Fleets, 1975 and 1979 (percent of grt)

Size Class (grt)	World Fleet 1975	Norway 1975	World Fleet 1979	Norway 1979
Below 6000	13	5	13	5
6000-10,000	12	4	10	2
10,000-20,000	19	13	19	9
20,000-40,000	17	15	15	12
40,000-50,000	5	8	5	4
50,000-100,000	14	24	15	25
Over 100,000	20	31	23	43
Average size	10,210 grt	22,550 grt	11,020 grt	24,710 grt

Source: Lloyd's Register of Shipping, reprinted in Norges Rederforbund, *Momenter til belysning av norsk og internasjonal skipsfart [Aspects Illuminating Norwegian and International Shipping]* (Oslo, 1976), 14; and *Momenter til belysning av norsk og internasjonal skipsfart* (Oslo, 1979), 14.

Investments in large ships are generally riskier than those in smaller vessels, but the increased risk has usually been rewarded through higher profits. Two factors can account for the higher risk: the lack of alternative employment for larger vessels and different demand patterns for ships in various

---

[12]If statistics including smaller vessels are used, the average size of the ships in the Norwegian fleet was more than twice as large as the Rest of the World.

size classes and/or market segments. The risk of operating a 100,000-dwt vessel is higher than that of operating four 25,000-dwt ships. Given this, Peter Lorange and Victor Norman note that "it is somewhat surprising that individual shipowners have chosen to invest heavily in supertankers."[13]

The type and size of the vessels clearly demonstrate the structural differences between the Norwegian and foreign fleets. The third element that distinguished the Norwegian fleet was propulsion. Due to the increase in bunker prices, turbine-driven vessels became less competitive than those powered by diesel motors. Turbine tankers were unfavourably affected by the changes in relative costs because they were constructed for a market with a totally different set of relative prices for inputs than prevailed after 1973.[14] A 50,000-dwt turbine-driven tanker used approximately eighty-five tons of "Bunker C" daily, whereas a similar-sized motor tanker used about fifty-five tons of low-grade diesel fuel.[15] Before the oil price hike, running costs were at approximately the same level, but after the increase there was a daily difference of US $1500 as a result of an increase in the price of bunkers of US $50 per ton and the additional thirty tons of bunkers needed to operate the turbine tanker.[16]

Vessels are generally laid-up when freight rates are insufficient to cover operating expenses less lay-up costs. The "reservation price" varies with the vessel's efficiency: efficient ships will accept lower rates than inefficient ones. Consequently, turbine tankers were among the first to be laid-up. The ships that had been among the most profitable when demand was high and increasing turned out to be particularly ill-suited to changes in the price of inputs. After the second oil price increase in the late 1970s, the economic disadvantage of a turbine-driven vessel increased to US $1.5 million annually.[17] But this was to some extent offset by a further fall in average vessel speed. The

---

[13]Peter Lorange and Victor D. Norman, "Risk Preference in Scandinavian Shipping," *Applied Economics*, V, No. 1 (1973), 49.

[14]In 1972, the cost composition for a new 100,000-dwt diesel vessel was as follows: fuel, ten percent; other operating costs, thirty-five percent; and financing costs, fifty-five percent. Seven years later the composition was forty-three, nineteen and thirty-eight percent, respectively; see Norway, Parliament, Stortingsmelding No. 52 (1980-1981), "Om skipsfartsnæringen" ["On the Shipping Industry"], 3.

[15]The figures in the example are taken from Roy Nersesian, *Ships and Shipping: A Comprehensive Guide* (Tulsa, OK, 1981), 29.

[16]Some shipping companies even chose to convert their ships from turbine to diesel engines; Norway, Stortingsmelding No. 52 (1980-1981), 26.

[17]Nersesian, *Ships and Shipping*, 29.

average recorded speed of Ultra Large Crude Carriers (ULCCs) was 9.5 knots in 1982 compared with fourteen knots in 1973.[18]

In the first period after the oil price increase, the Norwegian share of turbine-driven vessels was lower than the international average. Accordingly, Norwegian owners seemed to be in a better position with regard to fuel economy than their foreign competitors. But Norwegian owners had ordered a large number of mammoth, turbine-driven tankers. The share of turbine vessels in the Norwegian fleet increased considerably, from 30.7 percent in 1975 to 38.2 percent three years later. Over the same period, the share of turbine-driven vessels in the world fleet fell from 35.5 to 33.3 percent.[19]

The last element distinguishing the Norwegian from the world fleet was the average age of the vessels. The shipping market consists of both new, capital-intensive ships, and older, relatively labour-intensive vessels. The newer ships typically had a small crew and thus afforded marginal savings in crew costs, whereas older ships had lower financing costs. The Norwegian share of tonnage on order in the period 1950-1979 was generally higher than its share of the world fleet.[20]

The average age of the Norwegian fleet was six and one-half years in 1968, whereas the world average was eleven years.[21] In 1977, when the average age of Norwegian tankers was five years, the corresponding figure for the world tanker fleet was nine years. A year earlier, eighty-three percent of the Norwegian fleet was less than ten years old, compared to sixty-five percent of the world fleet and forty-four and five percent for the Greek and Cypriot fleets, respectively.[22] Some nations had tanker fleets with an age composition similar to Norway, and in 1974 Denmark, Sweden and Spain had even higher shares of modern ships. But all were much smaller than the Norwegian fleet; in 1974 each was two million grt compared to twelve million for Norway.[23]

---

[18]Figures for 1982 are from H.P. Drewry Shipping Consultants Ltd., *The Drewry Tanker Market Review* (London, 1982), table 12; and for 1973 from Drewry, *The Trading Outlook for Very Large Tankers* (London, 1975), 16.

[19]This is based upon grt figures from Norway, Statistisk Sentralbyrå, *Sjøtransport 1983* [*Maritime Statistics 1983*] (Oslo, 1984), 45 and 102.

[20]Norges Rederforbund, *Momenter til belysning*, 13. The only exception was 1968, when the Norwegian share of the world fleet was larger than its share of contracts.

[21]Fearnley and Egers Chartering Co. Ltd., *Review 1968* (Oslo, 1969), 12.

[22]OECD, *Maritime Transport 1976*, 128.

[23]The Japanese fleet was also more modern than the Norwegian fleet, but a considerable share of Japanese tonnage was owned by companies whose main interests

The structure of the Norwegian fleet was therefore distinct in an international perspective. There was a disproportionate share of tankers and bulk carriers, and Norwegians owned relatively large ships. The Norwegian fleet was also relatively modern and after 1976 had a higher share of turbine-driven vessels than the international fleet.

Even though shipping demand is international, and several of the most important factors of production are generally purchased in the international market, shipowners in different countries do not evaluate investments in ships and their operations in the same manner. Although there is an international maritime labour market, domestic legislation has impeded the use of foreign seamen in the fleets of most Traditional Maritime Nations (TMNs). This implies that domestic wages and manning regulations affect operational costs. Similarly, while debt financing can be acquired in the international market, equity generally comes from domestic sources. The alternative value of domestic capital is therefore important in determining the level of investment.

In a lecture in 1972, Professor Arnljot Strømme Svendsen presented three "survival mechanisms" for shipowners in "high-variable cost countries," such as Norway. The first strategy is risk-proneness, which Strømme Svendsen claimed was necessary given the cost advantages of low-wage countries. The second is "the continual ordering of new, modern vessels with technological superiority and operating advantages." This strategy should be combined with advantageous contracting and sales in a manner resembling an "asset-play" approach. The third survival mechanism is the development of new markets in which the shipowner might enjoy some kind of protection for a period.[24]

Although the use of the term survival mechanism may seem ironic when the decline of the Norwegian fleet is being considered, all three strategies were evident in Norwegian shipping. Since our focus is the crisis rather than the development of new and profitable market segments, the first two strategies, which clearly had negative implications, will be analysed here. Still, several Norwegian companies succeeded in establishing important positions in profitable niches, in stark contrast to the general economic performance of Norwegian and international shipping companies. For example, in the early 1980s Norwegian shipowners owned the world's largest fleet of cruise ships and were also prominent in chemical tankers and supply vessels.

The fact that Norwegians owned a disproportionate share of the largest and most modern vessels may be explained by the relative price of capital and labour in Norway. Due to the relatively high wage levels, Norwegian owners found it difficult to compete in the operation of labour-intensive ves-

---

were outside the shipping sector; this is an additional reason for the surprisingly low Japanese lay-up rates.

[24]Arnljot Strømme Svendsen, *Prospective Market Trends in the Shipping Trade* (Bergen, 1973), 13.

sels. Yet the relative abundance of capital available for shipping investments gave Norwegians an advantage in the operation of large, technologically-advanced vessels. The fleet structure was thus partly a result of cost differences and comparative advantage.

By the beginning of the 1970s Norway had developed into a high-cost country, with higher wages at prevailing exchange rates than other industrialised countries. Moreover, the average wage in shipping was considerably higher than in manufacturing. The relatively high cost of Norwegian seamen made it difficult for shipowners to compete successfully if labour-intensive vessels were employed, as labour costs comprised a relatively large share of total costs. Moreover, domestic regulations made the use of less expensive foreign labour difficult. Accordingly, labour-intensive vessels were not an alternative for Norwegian shipowners, who had to choose between leaving shipping or operating in the capital-intensive end of the market.[25]

If high wages made labour-intensive shipping unprofitable, the relatively low price of Norwegian capital made more capital-intensive vessels a realistic alternative. Investments in modern tonnage were facilitated by the relatively low alternative value of Norwegian capital. It is important to keep in mind that there was a considerably lower degree of international capital mobility in the 1960s and 1970s than today. Consequently, shipowners in different countries did not have identical investment opportunities – the alternatives to shipping investments were generally confined to the home country.

Comparisons of the return on capital are difficult and fraught with uncertainties. Analyses of the return on capital invested in shipping are particularly difficult due to the fluctuations in the freight market, the heterogeneity of the industry and the problems in estimating depreciation. Nonetheless, the capital-intensive nature of the industry makes investment decisions particularly significant.

Investments in Norwegian shipping were motivated by the fact that returns compared favourably with the alternatives. Ib Eriksen and Victor Norman have analysed the return from labour and capital invested in shipping during a "typical business cycle" in the period 1963-1972. Their conclusion was that the shipping industry provided returns at least as high as alternative investments.[26] A similar analysis, presented in a Norwegian White Paper, compared the gross return on capital invested in shipping with that on investments

---

[25]There were some Norwegian shipping companies operating in labour-intensive segments, particularly in regional Asian trades where manning requirements were waived, thus facilitating the operation of smaller and older vessels. Nevertheless, the share of Norwegian shipping in these trades was negligible.

[26]Ib Erik Eriksen and Victor D. Norman, "Skipsfarten i norsk samfunnsøkonomi" ["Shipping in the Norwegian Economy"], *Statsøkonomisk Tidsskrift*, LXXXVII, No. 3 (1973), 141.

in some of the most important capital-intensive manufacturing enterprises.[27] The comparison indicated that in five of the six years between 1968 and 1973, the return from shipping was higher, although not dramatically so. But the use of gross returns is to some extent misleading due to the higher depreciation in shipping. Still, the fact that shipping investments compared favourably compared with domestic alternatives can explain why a considerable share of Norwegian resources was directed into this sector. Moreover, low alternative returns on capital can explain both why Norwegian shipowners concentrated on the capital-intensive part of the shipping industry and why the Norwegian shipping sector was large in an international perspective.

International comparisons of the return on capital are subject to many of the same uncertainties as domestic assessments. Yet the variations in the most comprehensive study of the return on capital in the 1970s were so large that statistical differences are unlikely to account for all the discrepancies. The study, funded by the OECD, revealed large international variations. Table 6.4 shows that the return on manufacturing investments was considerably lower in Norway than in other OECD countries, except for the United Kingdom. Accordingly, shipping investments were relatively favourable for Norwegian investors since the alternative returns would have been low.

Table 6.4
Return on Manufacturing Investments, 1970-1979 (percent)

Year	Norway	Canada	US	France	Germany	UK
1970	10	14	20	22	21	10
1971	8	14	21	21	18	10
1972	10	16	24	21	16	10
1973	11	18	24	20	16	10
1974	13	19	17	23	16	6
1975	11	14	18	12	14	5
1976	9	13	22	11	16	5
1977	6	12	24	13	17	7
1978	5	13	23	13	17	7
1979	11	17	20	...	...	5
Average 1970-73	9.75	15.5	22.25	21	17.75	10
Average 1970-79	9.4	15	21.3	17.33	16.78	7.5

Note: Since Norway was not included in this study, it is likely that the figures here are estimates by the Ministry of Trade and Industry.

Source: Norges Offentlige Utredninger (1983:7), *Skipsfartens konkurranseevne* [*The Competitiveness of Shipping*], 66, which refers to OECD, *Profits and Rates of Return* (Paris, 1979).

---

[27]Norway, Parliament, Stortingsmelding No. 23 (1975-1976), "Om sjøfolkenes forhold og skipsfartens plass i samfunnet" ["On the Conditions for Seamen and Shipping's Role in the Society"], 63.

The high wage level and low alternative value of capital are important in explaining why Norwegian shipowners invested in modern, large bulk vessels. But these investments resulted in a low share of liner tonnage by international standards. This might seem strange, given that the liner segment is another capital-intensive part of the industry. Two factors can account for the relatively small Norwegian liner activity. First, Norwegian shipping companies have traditionally been small, taking advantage of the fact that economies of scale in most segments are related to the size of the ship rather than the size of the company. Economies of scale in the liner sector, however, are mainly connected to company size and the magnitude and frequency of services offered. Moreover, these aspects became more important with the onset of containerisation, which accelerated from the late 1960s.[28] Second, the links to the foreign trade of the home country are stronger for liners than for bulk carriers. Of the ten largest liner companies at the beginning of the 1980s, the Danish Maersk Line was the only one without a strong home-market base. The low share of liner tonnage thus to some extent reflects the relatively small Norwegian share of world trade and protectionist measures that made it increasingly difficult for cross-traders to operate in the liner sector. An analysis undertaken by the Norwegian Shipowners' Association showed that the introduction of the UN Conference on Trade and Development (UNCTAD) Liner Code of Conduct could reduce the volume of Norwegian liner transport by more than half.[29]

It has been claimed that high Norwegian manning costs might have contributed to the high lay-up rates. During the boom in the late 1960s and early 1970s Norwegian shipowners were able to earn substantial profits even with their relatively high operating costs. The freight market collapse, however, led to a greater focus on costs, and high manning levels may have rendered Norwegian owners unable to compete at lower freight rates.

Helen Thanopoulou has claimed that the high Norwegian lay-up rates may "seem as a self-explanatory situation due to the...high Norwegian costs in the 1970s."[30] Two elements make this assertion relatively unlikely. First, as Thanopoulou notes, there was no correlation between manning costs and lay-up rates in other countries. Second, most laid-up Norwegian tonnage was mammoth tankers for which manning costs were unimportant. If relatively labour-intensive Norwegian tonnage had been laid-up, high manning costs may have provided an explanation. But for those actually laid-up, such costs were more or less insignificant. Explanations therefore must be sought elsewhere.

---

[28]See Frank Broeze, *The Globalisation of the Oceans: Containerisation from the 1950s to the Present* (St. John's, 2002).

[29]Norway, Stortingsmelding No. 23 (1975-1976), 31.

[30]Helen A. Thanopoulou, "What Price the Flag? The Terms of Competitiveness in Shipping," *Marine Policy*, XXII, Nos. 4-5 (1995), 361-362.

The analysis above, emphasising Norwegian factor prices, is the "positive" explanation for the Norwegian emphasis on large, modern bulk vessels. By focusing on high domestic wages and the inferiority of Norwegian investment alternatives, shipowners are partly acquitted of their own misfortune. An alternative explanation may be that although Norwegian owners were unable to invest in the most labour-intensive segments of the shipping industry, their attitude towards risk made them order more and larger tonnage than the structure of factor prices could justify. This claim implies that the first of Strømme Svendsen's survival mechanisms was followed, that is, that the Norwegian strategies were riskier than those of their competitors. The question of risk is complex, and an evaluation of the relative importance of the attitude towards risk for the Norwegian fleet structure is impossible. Still, an assessment of the chartering policy may go some way towards describing the willingness of Norwegian shipowners to take risks.

With regard to chartering, the most important decision for shipowners is whether to tie their vessels to long-term charters or to operate them in the spot market.[31] Owners who operate mainly in the spot market have commonly been regarded as more risk-prone than those who tie their ships to longer charters. In this respect, chartering policy may be seen as an expression of the shipowners' attitude towards risk. The situation is more complex, however, and it is necessary to take into account the owner's assessments of future market developments when evaluating chartering policies.

In connection with the shipping crisis, there initially were large differences in the economic performance of companies that had secured their income through long-term contracts and those that had to offer their ships in a depressed spot market. There was no uniform behaviour among Norwegian tanker owners when it came to chartering policy – some shipowners had all their ships tied to long-term charters, others offered all their ships in the spot market and some chose a combination of the two. Despite the fact that some owners operated their ships in a risk-averse manner, in general Norwegian tanker owners were more exposed to market fluctuations than most of their foreign competitors. But the widespread portrayal of Norwegian shipowners as risk-seeking gamblers should be modified.

A commonly cited paragraph from an article in the American business publication *Fortune* can illustrate this. According to the article, "Norwegians own less than 15 per cent of the world tanker fleet, but account for nearly 50 per cent of the tonnage in the spot market."[32] Not only are the figures incor-

---

[31]The length and type of charters are of course also relevant. In connection with the oil price increase it was evident that some types of contracts were unprofitable to shipowners despite the fact that at the time of signing they seemed advantageous. The reason was the unexpected shift in exchange rates and bunker prices.

[32]"Betting $20 Billion on the Tanker Game," *Fortune* (August 1974), 124.

rect, but the assessment fails to consider an important reason for the Norwegian focus on the spot market: the lack of oil company-owned tonnage.

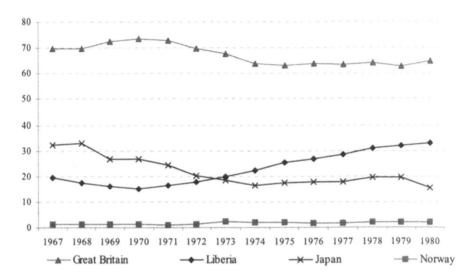

Figure 6.5: Share of Oil Company-owned Tanker Tonnage in Various Fleets, 1967-1980

Note: Measured in dwt. The totals on which the calculations are based deviate from those in the original source due to the subtraction of combination carriers, which were included in the total but not in the fleet by employment. Moreover, government and domestic tonnage has been excluded. In table II-12 in the original source the figures for the Norwegian and British fleets have been switched, but the figure has been corrected for this.

Source: Gilbert Jenkins, Martin Stopford and Cliff Tyler, *The Clarkson Oil Tanker Databook* (London, 1993), tables II-9 to II-22.

The uneven flag distribution of oil company-owned tonnage has generally been neglected in analyses of the Norwegian predicament. The fact that oil companies reserved cargoes for their own vessels implies that approximately a third of the tonnage in the tanker market essentially had secure employment (see figure 6.5). In 1978 the laid-up vessels owned by oil companies or their subsidiaries amounted to around one million dwt, representing less than one percent of their fleets.[33] The lay-up rates for the independent tanker fleet were therefore considerably higher than those for the aggregate fleet, as

---

[33]Figures from K. Dabrowski, "Comments on the Mechanisms of the World Shipping Market," *Maritime Policy and Management*, VIII, No. 2 (1981), 92.

most of the oil company-owned tonnage was insulated from the problem of securing employment. Yet oil company-owned tonnage in the Norwegian fleet was negligible.

If we adjust for oil company-owned tonnage, the difference between Norwegian and international lay-ups is less dramatic. By excluding oil company-owned tonnage, international lay-up rates increase by approximately fifty percent. A similar exercise for the Norwegian fleet would only have a minor effect, as most of the oil company-owned tankers on Norwegian register were oil barges or small tankers providing bunkers.[34] Adjusting for oil company-owned tonnage would lead to a 1.6 percent reduction in the size of the Norwegian fleet, whereas a corresponding correction for the world fleet would reduce its size by almost thirty-four percent.[35]

Measured as a share of the independent fleet, the international tanker lay-up rate was more than twenty-two percent in 1976, although tankers comprised only fourteen percent of the total fleet.[36] The first figure is more relevant in a comparison with Norwegian lay-up rates. The difference between the lay-up rates of the aggregate and independent fleets grew as oil companies increased their proportion of the world tanker fleet. The absence of oil company tonnage can only explain part of the problem of the Norwegian shipping industry, however. Even though it contributed to an explanation of the high Norwegian lay-up rates, it offered little consolation for shipowners unable to earn positive returns.

Another important reason for the high lay-ups was the manner in which Norwegian owners chose to operate their ships. Michael Porter has analysed Scandinavian tanker chartering policies as follows:

> The Scandinavians were considered gamblers among the gamblers of the shipping industry. They had placed more early orders for VLCCs and ULCCs than other nationalities and ran large portions of their fleet on the spot market. The Norwegians, for example, owned only 15 percent of the

---

[34]Nonetheless, Texaco Norway owned a fleet which in 1974 amounted to 372,000 dwt, comprising several medium-sized and one large tanker.

[35]In 1976, the only oil company-owned tankers of any size outside the Texaco fleet were the 32,000-dwt motor-tanker *Fjordshell* owned by AS Shellbåtene and the 18,500-dwt *Esso Slagen* owned by Norske Esso AS.

[36]This calculation is based on tonnage and lay-ups in dwt. The size of the independent and company-owned fleets comes from OECD, *Maritime Transport*, various years, while the figures for lay-up rates are taken from Fearnley and Eger, *Review*, various years.

world tankers, but accounted for nearly 50 percent of the tonnage on the spot market.[37]

The latter assertion – originating with the *Fortune* article and repeated in several places – needs to be modified.[38] According to figures from the British shipping analysts Clarkson, the Norwegian share of the world spot market reached its peak at the beginning of 1967 when more than one-quarter of the tonnage in this market was Norwegian. In the period 1974-1980, Norwegian vessels on average comprised approximately eighteen percent of spot market tonnage, a figure almost exactly twice as high as the average Norwegian share of the tanker fleet in the same period.[39]

In an historical context it is obvious that there had been a transformation of Norwegian chartering policies, though not necessarily one that is apparent from the distribution of the fleet between the various markets. The share of the Norwegian fleet on long charters decreased from the end of the 1960s onwards. Norman claims that "[i]n general, the paradox is that with the advent of supertankers, Norwegian ship owners have largely shifted from long-term timecharters to TCs of 1-3 years duration."[40] Yet when considering the share of Norwegian tonnage offered in the spot market, there appears to be no clear trend.

The reduced willingness to sign charters is not evident from the figures in table 6.5. The share of Norwegian tonnage on charters actually in-

---

[37]Michael Porter, "The Oil Tanker Shipping Industry," in Porter (ed.), *Cases in Competitive Strategy* (New York, 1983), 57. The claim is quoted in Stig Tenold, *Skipsfartskrisen og norske redere – en økonomisk-historisk studie 1973-1980* [*The Shipping Crisis and Norwegian Shipowners – An Economic-Historical Study 1973-1980*] (Bergen, 1995), 114.

[38]The claim that Norwegian vessels comprised fifty percent of spot-market tonnage is also advanced in Gunnar Nerheim and Bjørn S. Utne, *Under samme stjerne – Rederiet Peder Smedvig 1915-1990* [*Under the Star – The Shipowning Company Peder Smedvig 1915-1990*] (Stavanger, 1990), 237, where the authors refer to the above article in *Fortune*.

[39]These figures are based on dwt in Gilbert Jenkins, Martin Stopford and Cliff Tyler, *The Clarkson Oil Tanker Databook* (London, 1993), tables II-9 to II-22. The earlier restrictions apply. The figures group spot-market tonnage with tonnage in lay-up, tonnage used for storage and tankers used in the transport of grain. The high Norwegian share thus to some extent reflects the high proportion of Norwegian lay-ups, but this poses no big problem because the tonnage would have had to be offered in the spot market if rates had been sufficiently high.

[40]Victor D. Norman, *Norwegian Shipping in the National Economy* (Bergen, 1971), 64.

creased by more than thirty percent between 1966 and 1973. Two factors can account for this apparent deviation. First, the share of the fleet operating on charters tends to increase when freight rates rise. The two peaks in tanker freights in 1967 and 1970 can account for the historically high share of vessels on charters in the late 1960s and early 1970s. Charters entered into during the peak in 1967 may explain the high proportion of charters until 1970, whereas the peak in 1970 may explain the even higher share in the following years. Second, the figures do not reveal the length of the charters on which the vessels were operating. From the late 1960s Norwegian shipowners tended to reject long-term charters in favour of short- and medium-term agreements of one to five years.

Table 6.5
Share of the Norwegian Fleet on Charters, 1957-1979 (percent)

Time	Tankers	Tramp/bulk	Time	Tankers	Tramp/bulk
October 1957	86	64	July 1972	82	75
October 1960	71	49	July 1973	81	71
July 1963	59	33	July 1974	74	78
July 1966	60	48	July 1975	60	61
July 1969	67	51	July 1977	49	45
July 1970	65	60	July 1979	36	53
July 1971	87	77			

Source: Norway, Stortingsmelding No. 23 (1975-1976), 39; and Norges Rederforbund, *Momenter til belysning*.

While the figures showing the development of the various market segments can not directly substantiate the claim that Norwegians to a higher degree than previously rejected long charters, the high volume of Norwegian charters terminating in the first years after the freight market collapse indicates that there was a large share of short charters. Of the tonnage which had been chartered by July 1974, more than half operated on time charters which would be terminated by 1977 (see figure 6.6). Approximately two million dwt had secured charters into the first years of the 1980s, which represented less than five percent of existing Norwegian tonnage and contracts.

The steep reduction reflects the considerable number of charters that terminated in the period 1975-1979. Of the vessels on charter in July 1974, less than a quarter had been assured employment until the beginning of 1979. Accordingly, the ratio of short- and medium-term charters to long-term charters was approximately three-to-one. In an analysis of the plight of Norwegian shipowners, an international comparison of tonnage in the spot and charter markets will be more fruitful than an analysis of the historical development within the Norwegian industry. Again, there are difficulties in assessing the length of the charters because aggregate international statistics do not reveal charter length. Still, the specific chartering policy of Norwegians becomes relatively clear from analyses of the distribution of vessels between the spot and charter markets.

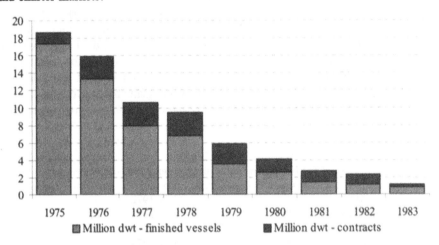

Figure 6.6:   Norwegian Charters as of 1 July 1974 and Development through 1983

Source: Archives of the Norwegian Shipowners' Association, folder 6 B K 74 – Krisen 1975 I, 010175-300675, "Various Information on the Economy of Norwegian Shipowning Companies," 30 June 1975.

The line in figure 6.7 reveals the extent to which the spot-market supply was composed of Norwegian shipowners. It is evident that Norwegian dominance decreased between 1967 and 1974. The line shows that the fifty percent Norwegian presence in the spot market, as claimed by *Fortune*, was simply wrong. Nonetheless, two elements in figure 6.7 require more explanation. The first is the proximity between the Norwegian and international spot market shares in the middle of 1974. This can be explained by two factors. One was the aforementioned correlation between freight rates and chartering – the low share of the Norwegian fleet in the spot market in 1974 reflects the high freight rate level in the latter part of 1973. A similar reduction was evident after the peak in 1970. Second, since the figures depict the share of the existing fleets offered in the spot market, the large number of Norwegian new building contracts without charters is not reflected in the figures. The second point that requires explanation is connected to the first. Why did Norway's share of the spot market increase much more rapidly than the rest of the world from 1975 onwards? The answer has already been given. One explanation is that a high share of Norwegian charters were short- and medium-term. Thus, the increase in the Norwegian share can be explained by the fact that a large portion of the charters terminated relatively shortly after the freight market collapse. Second, the Norwegian share increased when unfixed new buildings were delivered. Employment had not been secured for eighty percent of Norwegian new buildings, a proportion well above the world average.

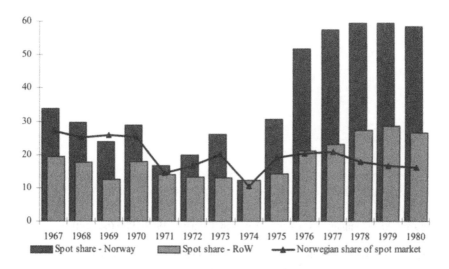

Figure 6.7: Spot Market Share, Norwegian and Rest of the World Tanker Fleets, 1967-1980

Note: The term "spot share" refers to all vessels which are not oil company-owned and which have not been assigned charters of more than three months, including ships used for storage and grain trading, as well as laid-up vessels.

Source: Jenkins, Stopford and Tyler, *Clarkson Oil Tanker Databook*, tables II-9 to II-22.

The share of Norwegian tanker tonnage that was inactive or operating in the spot market was more than twice as high as the rest of the world in every year from 1975. Figure 6.8 compares the employment pattern of the four largest tanker fleets in the period 1967-1980, depicting the share that had secured engagement either due to oil company ownership or chartering agreements. It shows that by July 1974 Norwegian shipowners had a smaller share of their fleet in the spot market than owners of Liberian-registered vessels. It also shows that the share of Norwegian tonnage on charter had increased by almost a third since 1967-1973. Thus, the situation in mid-1974 seemed favourable compared to owners with vessels registered in Liberia. Again, the two elements cloak the actual situation since differences in the length of charters and the entry of unfixed new buildings played important roles.

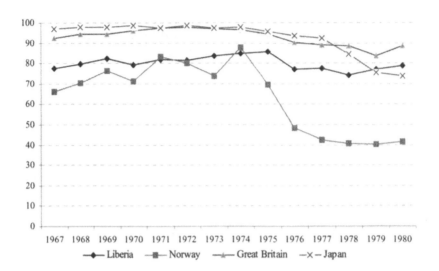

Figure 6.8: Company-owned and Chartered Tonnage, 1967-1980 (percent)

Note: Dwt. The lines refer to tonnage owned either by the oil companies or
               chartered for longer than three months. The Norwegian development is
               less dramatic than figure 6.6 suggests because this figure also considers
               charters signed after 1 July 1974.

Source: See figure 6.7.

Even though Norwegian shipowners appear to have had a more fortunate starting point than their Liberian competitors, with twelve as opposed to
fifteen percent of the fleet in the spot market, developments between 1974 and
1980 suggest that appearances may be deceptive. Whereas the share of the
Liberian fleet on charter and owned by oil companies was relatively stable,
falling less than ten percent between 1974 and 1980, the Norwegian share
plummeted: by 1977 the amount of Norwegian tonnage that had secured employment outside the spot market had been halved.

These figures refer to charters longer than three months. If figures for
those of more than twelve months are used, the share of the chartered Norwegian fleet in 1974 falls from almost eighty-eight to seventy percent, indicating
that more than a fifth of Norwegian charters may have had a duration between
three and twelve months.[41]

----

[41]Figures for charters of more than one year in July 1974 come from NOrway, Stortingsmelding No. 52 (1980-1981), 17. The fact that some charters may have
had a duration of more than one year, but were terminated in the period January-July,
should be taken into account.

In a government White Paper, sharply rising costs and uncertain exchange rate policy were used to explain the change in Norwegian chartering policies.[42] Changes in exchange rates and costs may explain why Norwegian shipowners began to prefer shorter charters, but neither factor was exclusively a Norwegian phenomenon. Instead, in order to evaluate the chartering policies of Norwegian shipowners we need to know both their aims and market assessments. As for the former, we may assume that these were the same for shipowners of all nations: above all, to maximise profits. Although some owners may have had other goals, such as preserving the reputation and tradition of the company, continuing the development of traditional business areas or even becoming the world's largest tanker owner, the notion of profit-maximising individuals should give a fair representation of the motive. When it comes to their assessment of the market, there is reason to expect a higher degree of variance.

Norwegian expectations about the development of the shipping market may have differed from those of their competitors. If Norwegians had a more positive view they would have avoided longer charters because they expected demand and supply to develop in a manner justifying this strategy. It is reasonable to assume that a shipowner anticipating a demand surplus would expect this to be reflected in high spot-market rates. The chartering policy could then be considered rational given their expectations. The growth of the world tanker fleet in 1974 equalled the average annual demand growth in the period 1966-1973. If this demand growth could be maintained, it would not be particularly risky to operate vessels in the spot market or on short charters. Expectations about a potential demand surplus not reflected in long-term charter rates may then have been one factor explaining the large share of Norwegian vessels without long-term contracts.

There is little doubt that Norwegian owners had an extremely optimistic view of the recovery of tanker demand. This may be because to a larger extent than their competitors they relied upon historical growth patterns.[43] The strong growth in tanker demand in the 1960s was undoubtedly one of the main causes of their positive market assessment. Norwegian shipowners may thus have been bedazzled by the combination of strong historical growth, high freight rates, ready availability of financing and positive demand projections from the oil companies. Until the market collapsed, Norwegian owners had been highly rewarded for their expansive and optimistic attitude.

Early in 1973 the Norwegian business publication *Kapital* claimed that the high value of shares in one shipping company "can be explained by the good prospects in the tanker sector." The company had just ordered a

---

[42]Norway, Stortingsmelding No. 23 (1975-1976), 39.

[43]Cf. the discussion in chapter 3.

280,000-dwt turbine tanker at a Japanese yard without the security of a long-term charter.[44] In the final months before the oil price increase the publication claimed that Norwegian shipping shares were undervalued despite the fact that several companies had enjoyed share price increases of between 300 and 1000 percent during the past year.[45] *Kapital* concluded that "[i]n the long-term, tanker shipping shares have a large growth potential. This applies to for instance Waage I and II, Ruth [and] Hadrian." Three of these companies operated mainly in the spot market, and all had contracted for mammoth tankers without the security of long-term contracts.

A leading Norwegian shipping economist asserted in 1972 that "[i]nternational seaborne trade will probably continue to grow, by some eight percent a year or more."[46] A year earlier, the managing director of the Norwegian Shipowners' Association expected annual growth rates of between 6.5 and eight percent.[47] The actual figure for the period 1973-1979 was less than 1.5 percent, and at the trough in 1982, world seaborne trade was smaller than ten years earlier. Gloating is not the motivation for the inclusion of these figures. Instead, they have been included because they illustrate two important points: first, that Norwegian expectations about transport demand were high; and second, that unforeseen events in the Middle East and actual development of transport demand played havoc with these expectations.

Yet such optimism did not prevail everywhere in the Norwegian shipping community. The shipbrokers Fearnley and Eger in their review of the 1972 shipping scene emphasised that the increase in trading distances was approaching a ceiling and that future demand increases would have to rely on increased consumption.[48] Moreover, the shipowner Odd Gogstad in December 1972 warned of a potential tanker oversupply.[49] Judging by subsequent Norwegian contracting, such pessimism was not shared by tanker owners.

The great expectations of Norwegian shipowners may be contrasted with the scepticism felt elsewhere. In its annual review of maritime transport

---

[44]*Kapital*, No. 5 (1973), 31.

[45]*Ibid.*, No. 19 (1973), 29. There was a considerable correlation between the value of shipping shares and the trend in freight rates.

[46]Strømme Svendsen, *Prospective Market Trends*, 12.

[47]David Vikøren, "Den internasjonale situasjon for norsk skipsfart" ["The International Situation for Norwegian Shipping"], *Statsøkonomisk Tidsskrift*, LXXXV, No. 1 (1971), 32.

[48]Fearnley and Egers Chartering Co. Ltd., *Review 1972* (Oslo, 1973), 5.

[49]Review of the 1972 shipping scene in *Farmand*, No. 51 (1972), 154.

the OECD claimed in early 1973 that:

> unless...demand for tonnage substantially exceeds the esti-
> mates made [by shipbuilding associations], it seems probable
> that there will be a significant oversupply of shipping during
> the next few years, and that shipyard capacity will be suffi-
> cient to meet any calls that may be made upon it up to the end
> of the decade.[50]

Following this dire prophecy, both Norwegian and international tanker contracting reached record heights.

If Norwegian tanker owners had a more positive view of future demand and freight rates than their competitors, their chartering policy can be
considered economically rational. The fact that Norwegians had historically
achieved above-average revenues by operating in the spot market undoubtedly
influenced their assessment.[51] Relatively positive expectations about future
markets can therefore not be ruled out as an explanation of why Norwegian
shipowners were reluctant to sign long-term charters. But it is difficult to
prove that systematically erroneous expectations were more prevalent among
Norwegians than among their competitors without a detailed survey of expectations in a variety of countries.

An alternative explanation of Norwegian chartering policy may be that
their attitude towards risk differed from their international competitors. Such
differences may have two explanations. They might result from disparities in
the willingness to take risks; alternatively, they may stem from perceptual
variations of the risk associated with various market segments.

Several authors claim that Norwegian chartering policy is evidence of
the willingness to take risks.[52] Yet the assumption that chartering policy is directly associated with a willingness to take risks is incorrect as long as expectations about future market developments are unknown. At any rate, the perceived risk propensity among Norwegian shipowners is not a postwar phe-

---

[50]OECD, *Maritime Transport 1972* (Paris, 1973), paragraph 98.

[51]The fact that formal market analyses were only utilised to a limited degree
by Norwegian shipping companies is emphasised in Victor D. Norman, "Shipping
Problems – Has the Market Mechanism Failed?" *Norwegian Shipping News*, No. 7
(1976), 26.

[52]See Porter, "Oil Tanker Shipping Industry," 57. The notion of risk-loving
Norwegians can also be found in Nerheim and Utne, *Under samme stjerne*, 217; and
Vidar Hole, "The Biggest Gamblers: Structure and Strategy in Norwegian Oil Tanker
Shipping during the Era of Growth 1925-1973" (Unpublished høyere avdelings thesis,
Norwegian School of Economics and Business Administration, 1993).

nomenon. S.G. Sturmey claims that the success of Scandinavian shipowners in the interwar period was the result of the fact that "they were not inhibited by traditional attitudes and were more flexible and enterprising in seeking new trades and in adapting their shipping enterprises to seek opportunities."[53] But the tankers responsible for the interwar expansion of Norwegian shipping were often on long-term charters. The risk of such investments was associated with low returns and the fact that this market was relatively unknown, rather than the manner in which the ships were operated.

Table 6.6
Annual Rates of Return as a Percentage of New Building Prices, 1967-1976

Year	Spot	Charter, 1-3 Years	Charter, 3-7 Years
1967	46.7	13.0	12.7
1968	35.5	15.2	9.6
1969	18.8	20.1	13.3
1970	48.8	14.4	8.8
1971	36.3	34.9	9.8
1972	11.3	40.7	17.0
1973	77.2	18.1	18.4
1974	7.5	40.2	7.8
1975	-14.2	22.2	20.7
1976	-1.6	-6.0	25.6
Average	26.6	21.3	14.4
Standard Deviation	25.5	13.0	5.0

Note: Annual rates of return measured as revenue less operating costs under Norwegian flag as percentage of new building price for an 80,000-dwt tanker built in 1966-1967

Source: Victor D. Norman, "Market Strategies in Bulk Shipping," in Einar Hope (ed.), *Studies in Shipping Economics in Honour of Professor Arnljot Strømme Svendsen* (Oslo, 1981), 16.

Norwegian shipowners have traditionally been regarded as entrepreneurs, and their organisations have been small and dynamic, exploiting the fact that economies of scale originate with the vessel rather than the company. Entrepreneurial organisations are often associated with a willingness to take risks, and creativity and the ability to innovate have often been regarded as the most important aspects of Norwegian shipping.[54] The description of Norwegian chartering policy, particularly when viewed in combination with high lay-up rates, may leave the impression that Norwegian owners were either naïve op-

---

[53]S.G. Sturmey, *British Shipping and World Competition* (London, 1962), 94.

[54]Arnljot Strømme Svendsen, "Er rederbegavelsenes tid forbi?" ["Is the Era of Shipowner Talents Over?"], *Bedriftsøkonomen*, No. 7 (1981), 352.

timists or die-hard gamblers. Analyses show, however, that the spot-market focus could be justifiable if an owner believed that he would be compensated for his increased risk with a substantial premium. Table 6.6 shows the annual rate of return of various chartering policies and gives an indication of the extremely profitable conditions for owners choosing a spot-market strategy in the late 1960s and early 1970s. Annual profits in the spot market between 1967 and 1973 averaged almost forty percent of the new building price, compared with twenty-two and thirteen percent for medium- and long-term charters, respectively. Indeed, the average return from a vessel operating in the spot market was almost twice as high as that of a vessel on three-to-seven-year charters. Spot-market returns exceeded those from long-term charters in every year but one, and they were higher than the returns from medium-term charters in each year except 1969 and 1972. The years analysed in table 6.6 may give a somewhat misleading picture of the superiority of the spot-market alternative. First, the period contained three booms, which was more than usual for a ten-year period; this increased both the average return and the standard deviation in the spot market. Second, a combination of policies may be preferable to a strategy in which all vessels are operated in the same manner.

The higher expected returns from spot-market chartering reflected the fact that the oil companies perceived the costs associated with a lack of tonnage as high. Since transport costs constitute a relatively small share of the oil price at the point of consumption, the oil companies have a considerable willingness to pay for transport capacity in periods where they fear a tonnage shortage. Their willingness to pay is reflected in high spot rates in periods of high tonnage demand and ensuing high returns on vessels operating in the spot market.

Several Norwegians have claimed that operating ships in the spot market was associated with a smaller degree of risk than tying them to long-term charters.[55] There were two reasons for this assertion. The first was that the high profits in the spot market gave sufficient compensation for periods when it was depressed. The second was the uncertainty with regard to exchange rates and costs, which added an element of risk to long-term charters. This aspect became particularly important when the volatility of the international economy increased. Several shipowners who had committed vessels to long-term charters in the late 1960s were caught off guard by the decline in the value of important currencies and unanticipated cost increases. It is therefore possible that the question is about the perception of risk rather than about the

---

[55]See, for example, Erling Dekke Næss, "Tankfartens problemer og utsikter" ["The Problems and Prospects of Tanker Shipping"] (Kristofer Lehmkuhl Lecture, Norwegian School of Economics and Business Administration, 1965), 13; or Hilmar Reksten, *Opplevelser* [*Experiences*] (Oslo, 1979), 164.

willingness to take it. An internal memorandum presented by a committee of the Norwegian Shipowners' Association claimed that:

> [f]or several of the vessels now laid up it may apply that they would have given higher revenues to their owners by the combination of, e.g., a three-year voyage-charter at a good rate followed by several years of lay-up, than what they would have achieved by e.g. a 5-8 year time charter that would have secured stable, but lower revenues.[56]

The disagreement between two of the biggest Norwegian owners may illustrate the different views on the perils of the various chartering strategies. The business policy of Sigval Bergesen involved operating ships on long-term charters, and Bergesen even went as far as to upbraid shipowners who chose to operate in the spot market.[57] Hilmar Reksten, on the other hand, claimed that owners operating in the charter market were "cotters" and asserted that the risk associated with this market was higher than in the spot market.[58]

A high propensity to take risks, different assessments of risk than their competitors and more positive expectations are all elements that may contribute to an understanding of Norwegian chartering policy. Although the relative importance of the various factors is difficult to ascertain, the analysis indicates that all were present.

The attitude towards risk among Norwegian shipowners can be illustrated by a study from the early 1970s. Peter Lorange and Victor Norman interviewed seventeen Scandinavian tanker owners in mid-1970 and found that under assumptions of reasonable liquidity, risk-proneness was widespread in the Scandinavian shipping industry.[59] The situation before the freight market

---

[56]Archives of the Norwegian Shipowners' Association, folder 6 B K 75 V – Krisen 1975 - 011275-310376, "Assessment of the Current Shipping Situation," 7 January 1976.

[57]In *Kapital*, No. 16 (1974), 11, Bergesen called Hilmar Reksten "hazardous" and declared that he did not regret losing the peak rates in the spot market due to his insistence on long-term charters.

[58]Reksten used the word *husmenn* which was used to characterise poor tenant farmers with life tenure – see, for instance, Hilmar Reksten, "Noen ideer om konkurransevilje og risikomomentet under strukturendringene i norsk tankskipsfart" ["Some Ideas about Willingness to Compete and the Risk Aspect during the Structural Changes in Norwegian Tanker Shipping"] (Kristofer Lehmkuhl Lecture, Norwegian School of Economics and Business Administration, 1971), 6; or *Norwegian Shipping News*, No. 19 (1973), 630.

[59]Lorange and Norman, "Risk Preference in Scandinavian Shipping," 57.

collapse corresponded well with the type of situation in which Norwegian owners were willing to take high risks. The propensity to take risks may thus be important in explaining the focus on the spot market and short- and medium- term charters.

Shipowners' strategies were also expressed through the acquisition and disposal of tonnage. For most Norwegian owners, tonnage acquisition implied contracting rather than the purchase of second-hand vessels. The most important element in this context was the type and timing of new building contracts, but the choice of shipyard and country of production was also a result of strategic decisions. Contracting activity is particularly relevant due to its importance for the structural development of the fleet, which again indicates a long-term strategy. The structure of the fleet is altered by both acquisitions and disposal. Whereas the current composition of the fleet is determined by past strategies, contracting and sales reveal the kind of vessels the shipowner intends to operate in the future.

The Norwegian shipowner and economist Erling Dekke Næss has claimed that there were three main reasons for the large number of tanker new building contracts prior to the shipping crisis.[60] The first was that shipowners lacked information about the mechanisms in the shipping market. Accordingly, decisions to order new tonnage failed to consider fully probable future demand. The second reason was that the behaviour of shipowners is often "sheep-like" in that some more or less blindly imitate the actions of their competitors. This pushed aggregate contracting to a level that could not be justified by realistic projections of demand. One of the causes of this herd mentality might have been signals from the charter market – owners made similar decisions because they interpreted signals in the same manner. Another reason may be that shipowners fear having to pay higher prices for new buildings if they enter a contracting boom at a late stage. The third reason was the tax system, which was of particular importance for Norwegian owners. Næss claimed that the depreciation rules, particularly in Scandinavia, facilitated continuous reinvestment in ships for tax purposes. He claimed that:

> with regard to the ship owners' contracting without taking the demand into account, the tax policies of several countries have lent a helping hand to the shipyards...[I]t may be worthwhile to take the risk of losses by speculating available capital on new tonnage if the alternative is to pay most of it as tax.[61]

---

[60]*Norwegian Shipping News*, Nos. 12/13 (1975), 13-21.

[61]Næss, "Tankfartens problemer," 11.

The article in which Næss is quoted also presents three additional factors that may have led to the large level of contracting. The first was the role of ship-brokers, who may have been more interested in their own short-term commissions than the long-term needs of their clients. The other factors were the lack of caution in the shipbuilding industry and the loan policies of the banks and investment syndicates.

The contracting by Norwegian shipowners before the shipping crisis was influenced by all the elements described by Næss. Three aspects of Norwegian contracting differed from the world norm. The first was the type of vessels ordered. The second was the level of contracts relative to the size of the existing fleet. Both reflect aspects that were analysed in connection with the fleet structure. The last characteristic was the amount of tonnage that had been ordered without transport assignments having been secured.

The contracting in 1973 partly reflected the preference of Norwegian shipowners for large tankers and dry bulk vessels. Still, in a comparison of Norwegian and international statistics, the differences between Norway and other countries were less conspicuous with regard to new building contracts than with regard to the composition of the existing fleets. The relative similarity with regard to contracting was partly because 1973 was extraordinary in that the amount of tanker tonnage contracted at the international level was considerably higher than in more "normal" years (see table 6.7). At the international level, tanker contracts constituted more than eighty percent of the total tonnage ordered, compared with an average of approximately sixty percent in the period 1968-1972. Compared with the existing fleets at the time, it is evident that the international orders had acquired a "Norwegian" character. The high share of tankers ordered and the relative absence of general cargo vessels imply that international contracting was more similar to the structure of the Norwegian than the international fleet at the time. Moreover, the considerable difference in average size was hardly reflected in the contracting. The difference in average size between Norwegian and international tanker orders – 185,000 dwt versus 171,000 dwt – was considerably smaller than in the case of the existing tanker fleets – 96,000 dwt versus 63,000 dwt.

Although the structural composition of the 1973 contracting was relatively similar between Norwegian and international shipowners, the level as a share of the existing fleet differed. Norwegian shipowners accounted for fourteen percent of world tanker contracting, whereas their share of the tanker fleet was approximately ten percent. The disproportionate contracting was the result of Norwegian cost structures and the fact that Norway had a faster replacement of ships than other countries.[62]

---

[62]Cf. the discussion of the Norwegian fleet structure in Norway, Stortingsmelding No. 52 (1980-1981), 18.

Table 6.7
Norwegian and International Contracting and Order Books, 1973 (dwt)

	% of Norwegian Contracting 1973	% of International Contracting 1973	Average Size – Norway	Average Size – International
Tankers	84.0	80.5	184,974	171,365
Combination	1.4	2.9	121,462	137,753
Dry bulk	14.3	11.2	72,348	44,438
Others	0.4	5.4	5714	9561

Note: Contracting figures refer to new buildings ordered in 1973, whereas figures for the average size of vessels refer to the order book on 1 January 1974. With regard to average size, the international figures refer to bulk vessels above 10,000 dwt, whereas the Norwegian figures in the original source refer to vessels larger than 1000 dwt. This has been adjusted by means of the size distribution of Norwegian orders in Fearnley and Egers Chartering Co. Ltd., *Review 1973* (Oslo, 1974), table 25.

Source: Fearnley and Eger, *Review 1973*.

It is possible to explain a large share of the variations in Norwegian tanker contracting by means of the rate level for medium-term charters. Siri Pettersen Strandenes claimed that Norwegian owners were more influenced by variations in the current freight level than were their foreign colleagues.[63] A given increase in the period rate led to an eight percent increase in Norwegian contracting but only a five percent increase internationally. The high level of Norwegian contracting in 1973, triggered by the high rates, haunted owners for a considerable time after the freight market collapse and was one of the reasons that the owners were particularly hard hit by the crisis.

Figure 6.9 shows the violent fluctuations in the Norway's share of total tanker contracting.[64] In 1970 and 1973, when freight rates were high, the Norwegian share was in the region of fourteen percent. It was about the same in 1972, when the rate level was low. Due to the rapid changes in freight rates, however, annual figures may be somewhat misleading. The majority of the 1972 contracts were signed in the last part of the year, after the market had shown signs of recovery. The tonnage ordered by Norwegian owners at foreign yards increased from 304,000 grt in the third quarter to almost 3.2 mil-

---

[63]*Norwegian Shipping News*, Nos. 13/14 (1979), 26.

[64]The large increase in Norwegian contracting in 1977 was largely the result of two ULCCs contracted by Bergesen at the Japanese Mitsui yard which accounted for more than two-thirds of Norwegian contracting for the year. The low level of international contracting at this time gave Norway a large share of total contracting. See *Norwegian Shipping News*, No. 8B (1977), 34-35, where the company's reason for contracting in a generally "dead" new building market was explained.

lion in the fourth quarter.[65] The Norwegian share of international tanker contracting in the depressed market in 1971, when new building prices were low, was only around five percent. The low figures in 1974 and 1975 show the relatively quick response to the changed state of the market. Indeed, if cancellations are deducted from new orders, net contracting was negative. In the third quarter of 1974 the tonnage cancelled was higher than the new tonnage contracted, and this situation persisted, with one exception, until the third quarter of 1976.[66]

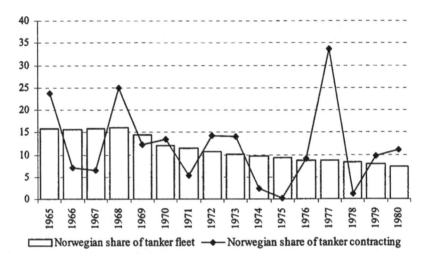

Figure 6.9: Norwegian Share of the Tanker Fleet and Contracting, 1965-1980

Note: The figures are to some extent affected by the fact that the statistical basis differs – the figures for international tanker contracting refer to vessels above 10,000 dwt, while the Norwegian figures include vessels in the 1000-10,000 dwt range. An adjustment would not lead to significantly different results.

Source: Fearnley and Egers Chartering Co. Ltd., *Review 1976* (Oslo, 1977); and *Review 1981* (Oslo, 1982), tables 3, 20 and 27.

The tonnage ordered in 1974 and 1975 was mainly smaller vessels. The average size of Norwegian tankers ordered in 1973 was 184,000 dwt, re-

---

[65]In the third quarter, 290,000 grt of tanker tonnage was ordered, compared with 2,674,000 grt in the fourth quarter. Norway, Statistisk Sentralbyrå, *Statistisk Månedshefte* [*Monthly Bulletin of Statistics*] (Oslo, November 1972 and April 1973), table 13.

[66]*Ibid.*, various years.

flecting the fact that fifty-two vessels of more than 100,000 dwt had been ordered. In 1974 no orders for vessels of more than 60,000 dwt were placed by Norwegians. The average size fell to 30,000 dwt in 1974 and further to 4000 dwt in 1975. The high Norwegian share of international contracting in 1977 was mainly the result of contracts for Bergesen's two large tankers and reflected the cessation of international tanker contracting. Whereas Norwegian orders in 1977 comprised more than a third of the international total, they would have constituted less than one percent of international contracts in 1973.

The analysis of Norwegian ordering discloses two features that we would expect, given the analysis of the fleet structure. It was dominated by large bulk vessels, and the Norwegian share of international contracting was high relative to its share of the world fleet. The relatively small differences in the structural composition of Norwegian and international ordering in 1973 must be understood in terms of the market at the time. The basis was that orders by foreign shipowners had a "Norwegian" flavour, not *vice versa*. With regard to vessels on order, Norwegian tanker and combination carrier orders corresponded to 106 percent of their current fleet, whereas the corresponding figure for international owners was eighty-two percent.

The structural differences between Norwegian and international contracting in 1973 were too small to explain fully why Norwegian owners were affected relatively severely by the crisis. As for the level of contracting, Norwegian owners were more eager in the new building market than their international competitors, a fact that to some extent can explain why they were hit harder by the freight market collapse. But the single most important difference was neither structure nor level, but rather the amount of new buildings that had been ordered without employment being secured. These unfixed new building contracts were based on expectations about continuous growth in demand for tanker transport and were an important reason for the Norwegian hardship. After the oil price increase, owners who had ordered unfixed vessels when rates were high had to choose between receiving vessels that had to operate in a market with surplus capacity or paying large cancellation or conversion fees to the shipyards. Appendix 2 lists Norwegian tankers and combination carriers above 200,000 dwt on order in the beginning of 1974. The vessels had been ordered by twenty-five different shipowners, and twenty-one of these had ordered vessels for which charters had not been secured. The close relationship between various Norwegian shipowners implies that several companies may have participated in the investments without being included in the list.

A considerable share of the vessels presented in the appendix were cancelled. Of the sixty new buildings on order, thirteen had been fixed on time charters, ranging in duration from two to eight years. Figure 6.10 shows the amount of unfixed tonnage, as well as the amount that had secured employment in the period 1976-1983. The bars for the various years refer to the amount of tonnage on charter at the beginning of the year.

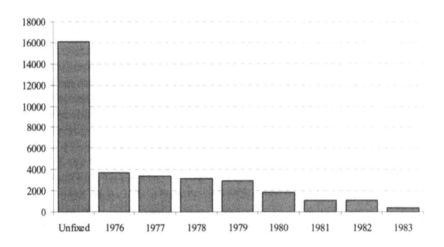

Figure 6.10 Norwegian New Buildings on Charter, 1976-1983 ('000 dwt)

Source: See appendix 2.

        Norwegian contracting in the early 1970s was important in explaining
why Norwegian shipowners were so affected by the international shipping crisis. Compared with their foreign competitors, Norwegian shipowners did not
order more tonnage than they had traditionally done, but a large portion of the
contracts had been signed when the price of new buildings was high or increasing. Moreover, more than eighty percent of the largest tanker tonnage on order
had not secured employment. This is where Norwegian new building orders
really differed from those of their competitors. Of the non-Norwegian tanker
tonnage on order in January 1974, thirty-seven percent had secured time charters, thirty-two percent were oil company contracts and thirty-one percent were
intended for the spot market.[67] The figures refer to new building contracts for
which data on employment are available and are based on 398 observations for
the international and sixty for the Norwegian fleet. It is thus evident that
whereas the portion of the existing Norwegian fleet in the spot market corresponded to the international share, the situation was different with regard to
the "fleet" of new building contracts – in this respect Norwegian owners differed substantially from owners elsewhere.
        Richard Goss claims that the "real risk" of contracting large tankers
was only evident when the first time charter had expired, at which point most

---

        [67]Calculations based on figures from the Norwegian shipbroker Johan G. Ol-
sen.

of the debt had been repaid.[68] The new buildings of twenty-one Norwegian shipowners would not even reach this critical point, as they had not been fixed on charters that could secure these first instalments. The eager contracting in the early 1970s affected Norwegian shipowners in two ways. Several had to receive new buildings for which employment could not be found, and the laid-up vessels constituted a drain on their resources. Others avoided this by paying massive cancellation and conversion fees. It has been estimated that fees paid by Norwegian owners from 1975 to 1977 were in the region of NOK 1.5 to two billion.[69] By following a more prudent contracting strategy, Norwegian shipowners could have avoided paying a "penalty" of such proportions.

The heterogeneity of the Norwegian shipping community was emphasised earlier, and the high number of unfixed Norwegian contracts may have been the result of the influence of a few expansive and risk-seeking shipowners. If we adjust for the effects of the contracting by the three Norwegian shipowners usually presented as risk-lovers – Biørnstad, Waage and Reksten – and take into account the lack of oil company tonnage in the Norwegian fleet, the picture is partially modified. Yet significant differences between Norwegian shipowners and those in other countries still remain. On the international scene, forty-five percent of independent shipowners' new buildings were unfixed compared with seventy-three percent for Norwegian shipowners, if we exclude the seventeen unfixed vessels ordered by Biørnstad, Reksten and Waage. The difference is still substantial, and shows that the large amount of unfixed new buildings was not the result of the influence of a handful of gamblers but had a more fundamental basis in the Norwegian industry.

The Norwegian orders may partly be accounted for by the relative price of labour and capital in Norway. Shipowners were encouraged to invest in modern vessels that utilised economies of scale and the best available technology, simultaneously minimising manning costs. By ordering new, capital-intensive and technologically superior vessels, and disposing of these when even more advanced alternatives became available, Norwegian shipowners were able to utilise their comparative advantage with regard to capital costs and reduce the importance of their comparative disadvantage in manning costs. This strategy necessitated a high level of contracting.

Moreover, Norwegian shipowners' evaluation of future transport demand and their attitude towards risk may also have influenced the amount of new buildings. The mechanisms are the same as in connection with chartering

---

[68]Richard O. Goss, "A Comment on Risk Preference and Shipping Decisions," in Peter Lorange and Victor D. Norman (eds.), *Shipping Management* (Bergen, 1973), 185.

[69]Archives of the Norwegian Shipowners' Association, folder 6 B K 75 – Krisen 1975 VII, 011077-311277, memorandum to Hallvard Bakke, Norwegian Minister of Trade and Shipping, 10 October 1977.

strategies. If Norwegian owners were more optimistic about future transport needs than their international competitors, this would have been reflected in a larger share of tonnage on order. If they intended to benefit from the peaks in transport demand, and were willing to suffer the anguish of oversupply, we should also expect high contracting levels.

Berg Andreassen has shown how investment levels were influenced by risk preferences.[70] Shipowners willing to take risks will invest in a manner that ensures that the peak transport demand will be met. Risk-averse shipowners, on the other hand, will typically base their decisions on more conservative estimates of future transport demand. In this connection, it is worth noting that the Norwegian behaviour corresponded closely to the strategy that Berg Andreassen claims "risk lovers" will follow in prosperous times. Yet it is not necessarily true that this behaviour was the result of a willingness to take risks; the institutional setting may also affect the actions. More specifically, the manner in which the tax system is designed may influence strategic decisions, implying that after-tax evaluations may differ from pre-tax prognoses. As Lewis Fischer and Helge Nordvik have noted, "the principal method of subsidising international shipping since the Second World War has been through manipulations of domestic tax systems."[71]

Due to the fact that shipping is a mobile sector in which a large share of the factors of production are bought and sold internationally, it is necessary to take special precautions in connection with the taxation of shipowning companies. As a result, the tax treatment of Norwegian shipowners has traditionally been regarded as favourable compared with other sectors of the economy.[72] The relatively dominant position of shipping in the Norwegian economy can partly be explained by the fact that the tax system has been preferential for shipping investments, thus constituting an indirect subsidy relative to other sectors. Tor Wergeland has pointed out four areas in which the tax system has been particularly preferential: liberal depreciation rules that have made large

---

[70]J.A. Berg Andreassen, "Risk and Investment Decisions in Non-liner Shipping," *Maritime Policy and Management*, XVII, No. 1 (1991), 22-30.

[71]Lewis R. Fischer and Helge W. Nordvik, "Subsidy and Protection in National Shipping Industries around the North Sea since World War II," in Randi Erstevåg, David J. Starkey and Anne Tove Austbø (eds.), *Maritime Industries and Public Intervention* (Stavanger, 1995), 89.

[72]Terje Hansen, *Aarbakke-utvalgets skisse til reform av bedrifts og kapitalbeskatningen og norsk skipsfartsnæring* [*The Aarbakke Committee's Sketch for Reform of Company and Capital Taxation and the Norwegian Shipping Industry*] (Bergen, 1989), 45. Despite the favourable treatment, Norwegian shipowners called for relief in connection with the shipping crisis; see Norges Rederforbund, *Årsberetning 1978/1979* [*Annual Report*] (Oslo, 1979), 5.

tax deductions possible; the fact that losses can be carried forward and deducted from future profits; the "gross principle" with regard to taxation, implying that the purchase price is the basis for depreciation when second-hand tonnage is bought; and the fact that sales profits can be written off against new investments.[73] As a result of the favourable treatment of shipping investments, the manner in which owners employed their capital has been affected. The special tax treatment of shipping has been motivated by the fact that its assets are far more costly than those in other sectors, and that a higher turnover of assets is necessary.[74] Accordingly, the tax practices have been favourable in order to compensate for the rapid deterioration of shipping assets relative to other investments.

Although the special rules reflect actual features of the shipping industry, such as the relatively short lifespan of the assets and the limited geographical restrictions on the capital, the arrangements have substantial consequences for behaviour. The government, for instance, may through its depreciation allowances have influenced contracting decisions.[75] Indeed, the Norwegian authorities recognised this and admitted that "[t]he special rules for shipping taxation will lead to a considerable tax credit for companies with economic results which make the rules effective."[76]

The depreciation rate for Norwegian vessels varied from six to eight percent annually for tankers and most other vessels, and from five to seven percent for dry cargo vessels. But two elements are of particular importance with respect to depreciation: the access to *tilleggsavskrivninger* (additional depreciation) when the vessel is delivered, representing five percent of the original cost annually and fifteen percent in total, but limited to fifty percent of ordinary depreciation annually. Thus, if the shipowner chooses eight percent ordinary depreciation, additional depreciation may be four plus four plus four plus three percent and would end after four years. Alternatively, shipowners had access to *åpningsavskrivninger* (accelerated depreciation) of up to twenty-five percent of the original cost, but the accelerated depreciation could not exceed fifty percent of the taxable income of the shipowner. The most impor-

---

[73]Tor Wergeland, *Et konkurransedyktig Norge – Norsk skipsfarts konkurranseevne* [*A Competitive Norway – The Competitiveness of Norwegian Shipping*] (Bergen, 1992), 47.

[74]Norway, Stortingsmelding No. 52 (1980-1981), 23.

[75]See, for instance Næss, "Problems and Prospects," 11; Porter, "Oil Tanker Shipping Industry," 57; and Kim Dobrowen, *et al.*, *Rederi og kapitaltilførsel – Hovedproblemer i moderne rederifinansiering* [*Shipowner and Capital Supply – Main Problems in Modern Ship Financing*] (Oslo, 1981), 241.

[76]Norway, Stortingsmelding No. 23 (1975-1976), 65.

tant feature of the access to accelerated depreciation is that it could begin as soon as the first instalment on the building contract had been paid.[77]

The access to additional or accelerated depreciation shortened the period of ordinary depreciation and may have contributed to the liberal Norwegian depreciation rules for shipping investments. This has important ramifications. First, companies will prefer investments in capital assets to financial investments. Second, if the period of depreciation differs from the life of the asset, companies have an incentive to invest in the asset for which the depreciation allowances are favourable, in particular if additional or accelerated depreciation is possible. Third, companies operating in relatively capital-intensive sectors have an advantage compared with those in less capital-intensive sectors, and companies will substitute capital for labour. Finally, expanding companies are favoured because the tax credit that the depreciation represents makes it easier to finance increased capital needs.

The favourable Norwegian depreciation allowances may have induced shipowners to order new vessels without sufficient analysis of the market conditions at times when profits were large. The existence of anticipated depreciation on shipbuilding contracts meant that shipowners with large profits could reduce their tax burden considerably by ordering new ships. This may contribute to an explanation of the strong correlation between freight rate levels, Norwegian contracting and the apparent "herd behaviour."

Another element was the benefits of reinvestment in connection with revenues from the sale of assets. If a vessel were sold at a profit, the owner would receive a suspended tax relief on the capital gains. The tax on the profit could be deducted from the book value of a new investment within the following eight years. The writing down of sale profits when new investments were undertaken did not reduce the basis for depreciation and came on top of ordinary and additional or accelerated depreciation. The maximum depreciation, including accelerated depreciation, on a tanker meant that the vessel was fully depreciated after ten years. Yet the lifespan of the vessel was generally considerably longer, and its value therefore represented a "hidden reserve" for the shipowner.[78] A fully depreciated vessel would increase the shipowner's tax liabilities unless it was sold and the profits reinvested. Through continuous reinvestment, the temporary tax relief inherent in the depreciation rules and suspended tax on profits from sales became permanent.

Johan Seland of the Norwegian Shipowners' Association strongly refuted the assertion that the tax system explained the timing of Norwegian con-

---

[77]The Norwegian tax regime made provisions for accelerated depreciation in other sectors as well, although the deductible amount was smaller, and depreciation could begin only when the building of the asset had begun, not when the first instalment was paid.

[78]Dobrowen, *et al.*, *Rederi og kapitaltilførsel*, 238.

tracting. In an interview he claimed that:

> [v]ery few Norwegian new building contracts could not have been postponed for a considerable time if the purpose was to gain preferential depreciation. As the contracting has happened at an earlier time, and been larger than necessary to secure the depreciation, it is obvious that other factors have been more important for the decisions.[79]

His assertion is partly correct. The fact that capital gains from sales could be suspended from taxation for eight years gave shipowners limited incentives to invest in new tonnage immediately after vessels had been sold. But the issue of capital gains is not the most important, since although capital gains could be deferred, freight income could not. Indeed, the liberal depreciation rules only became important when the company earned a profit, and in particular when the profits were high enough to enable maximum depreciation. The fact that the tax burden would be high in years with high freight rates motivated owners to order new tonnage. If new contracts were placed and accelerated depreciation was activated, the tax liability was considerably reduced.

This analysis takes into account the two most important aspects of the Norwegian tax system and shows its design, in particular the depreciation rules, may have contributed to the high level of Norwegian contracting. Other aspects of the Norwegian tax system, such as the ability to carry losses forward and the provision for classification, devaluation and self-insurance funds, have not been included here, although they do distinguish the tax treatment of shipping from that of other domestic industries. The taxation practices were changed in the wake of the crisis, however. For instance, shipowners were granted a reduction on income equal to twenty-five percent of the price of vessels contracted at Norwegian yards between 15 February and 31 December 1977, a rule motivated by the predicament of shipbuilders rather than shipowners.[80]

The hardship for Norwegians after the shipping crisis was reflected both in high lay-up rates and a sharp reduction in the size of the Norwegian fleet. The three features analysed above – fleet structure, chartering strategy and contracting – explain why Norwegian shipowners were more affected by events than their international competitors. The analysis of fleet structure showed that Norwegian shipowners had invested a disproportionate share of their resources in the types and sizes of vessels which were most adversely affected by the crisis. Norwegians were thus more vulnerable to the changed

---

[79]*Kapital*, No. 3 (1975), 8-10.

[80]Dobrowen, *et al.*, *Rederi og kapitaltilførsel*, 244.

conditions in the market than foreign shipowners. The analysis of the chartering policy showed that while Norwegian and international shipowners had
similar proportions of tonnage in the spot market at the outbreak of the crisis,
this apparent similarity concealed two aspects that made Norwegians more
vulnerable. First, they had a relatively large share of short- and medium-term
charters, as shown by the large number that terminated in the period 1974-
1977. Second, although the share of the existing fleet on charters was about
the international norm, Norwegians had a considerably higher share of unfixed
new buildings. This was true in particular for large tankers, where eighty percent of the tonnage ordered was unfixed, compared with thirty-one percent for
foreign shipowners. It was these unfixed new building contracts, rather than
contracting *per se*, which differentiated Norwegian from foreign shipowners.
Given the existing structure of the Norwegian fleet, the effects of the Norwegian tax system and the influence of Norwegian factor prices, a relatively large
share of new buildings on Norwegian account should have been expected. But
while the relatively rapid replacement of tonnage provided Norway with a
comparative advantage in the 1960s and early 1970s, a large fleet of new
builds became a liability when there was a tonnage surplus.

The factors discussed above are paramount to an explanation of the
high lay-up rates and the economic difficulties of Norwegian shipowners. Yet
two other factors – exchange rates and costs – may also have affected the shipping sector negatively, without necessarily being visible through higher lay-up
rates. The large exchange rate fluctuations after the collapse of the international monetary system increased the risks entailed in contracting and chartering. The currency instability may have had more severe implications for Norwegian shipowners than for foreigners for two reasons. First, the Norwegian
fleet was relatively modern, implying that a relatively small share of the debt
had been repaid. Norwegian shipowners thus had relatively high financiing
costs, which may have exposed them to exchange rate losses.[81] Second, the
Norwegian exchange rate balance – the relationship between the currency in
which income is derived and that in which costs are incurred – may have
added to the vulnerability of Norwegian shipowners.

The hardship may also have been affected by cost trends, which have
been used to explain why Norwegian owners operated such a large share of
their ships in the spot market. A prominent Norwegian shipowner has com-
mented that "ordering new buildings without the backing of freight contracts
was much less of a gamble than ordering on the basis of long-term charter par-

---

[81]In late 1976 the Norwegian debt on existing vessels and contracts amounted
to almost NOK thirty-six billion (US seven billion), of which four-fifths was related to
foreign creditors. Six months earlier the debt on the world fleet and contracts was estimated at US $34.7 billion. Norwegian shipowners were thus responsible for about
twenty percent of international shipping debt, whereas their share of the world fleet and
order book was slightly more than eight percent.

ties which give no security against rising costs and inflation."[82] The high inflation in the early 1970s led to a considerable increase in the cost of building and
operating vessels. Although this generally affected all shipowners, regardless
of nationality, there might be some support for the argument that Norwegian
owners were impacted more negatively because their aggressive ordering strategy forced them to pay large cancellation fees or to accept new buildings that
at the time of delivery were worth far less than the price the owners had
agreed to pay.

Norwegian owners may also have been adversely affected due to differences in the price of the factors of production, which often were higher in
Norway.[83] This, however, was offset to some extent by the international character of the shipping market, since most factors of production were bought in
an international market at prices which were generally equal for all buyers.
The only area in which national operating costs were of any particular significance was in the manning of ships. Here there was a problem for Norwegians,
since total manning costs for domestic shipowners more than doubled between
1971 and 1977. Indeed, a Norwegian White Paper claimed that the strong domestic price rises between 1974 and 1977 weakened the competitiveness of
Norwegian shipowners.[84] It appears, however, that this was not very significant in a long-term perspective because manning costs were of limited importance for the majority of Norwegian tonnage, and developments in the last part
of the decade compensated for any previous Norwegian cost disadvantage. In
fact, a 1983 survey of the competitiveness of Norwegian shipping concluded
that "the wage development of Norwegian seamen has been moderate in the
1970s."[85]

Finally, the analysis shows that the image of Norwegian shipowners
as "gamblers" requires modification. Given their environment, including tax
policies and factor prices, as well as expectations about continuing strong
transport demand growth, their investments were fairly rational. Indeed, judging by the expectations prevalent in 1973, the Norwegian strategy on fleet
structure, chartering and contracting to a large extent was economically sensi-

---

[82]See Reksten, "Noen ideer," 18. The quote is taken from an interview with
Reksten in *Norwegian Shipping News*, No. 19 (1973), 630.

[83]This implies that the price of one factor of production may develop differently for different shipowners. The pattern of the operating costs of Norwegian ships
illustrates to some extent that trends in factor prices varied between countries; see
Norway, Stortingsmelding No. 23 (1975-1976), 98.

[84]Norway, Stortingsmelding No. 52 (1980-1981), 18. See the index of costs
for Norwegian ships in Norges Rederforbund, *Momenter til belysning*, table 8.1.

[85]Norges Offentlige Utredninger (1983:7), 56.

ble. Yet when OPEC's policies and the subsequent freight market collapse caused the reality to diverge from expectations, Norwegian shipowners were more severely affected than those in other countries.

Chapter 7
Structural Transformations in Norwegian Shipping

The transformation of Norwegian shipping during the crisis can be analysed using information on the population of shipowners. The most important elements are the size and number of shipping companies; their spatial and business concentration; and the processes of growth and decline. The analysis is based on annual time series which span the period 1970-1987, enabling analysis of the dynamics and transformations during an era with both strong expansion and sharp contraction. The database, which includes all vessels above 5000 gross registered tons (grt) and all companies owning such vessels, is based on the Veritas register.[1]

There is a high degree of correspondence between the figures in the database and those published elsewhere. Figure 7.1 compares the data with other independent sources. The figures provided by the Organisation for Economic Co-operation and Development (OECD) are higher in the first part of the period and lower thereafter since they are based on mid-year data, whereas the database uses the start of the year. The OECD figures are thus likely to be higher in periods of expansion and lower when the fleet is shrinking.

Although the database corresponds well with other sources, this does not necessarily mean that it suits the objects of this analysis. One potentially problematic feature concerns the term "shipping companies." The vessels have been categorised according to their *korresponderende reder* (managing owner), but he was not necessarily the legal owner of all the vessels he managed.[2] Nevertheless, using a different organising principle would introduce a degree of arbitrariness. Moreover, it is logical to use the managing owner, since he made the decisions and formulated the strategies; hence, it is his behaviour in which we are primarily interested.

---

[1]Det Norske Veritas is the leading Norwegian ship classification agency. Some information problems, such as time lags and incomplete information, have been corrected insofar as possible using other sources. As the register is compiled by an independent agency, whose *raison d'être* is to provide accurate and reliable information to outside agents, there is no reason to expect any bias.

[2]For instance, in 1975 the Oslo-based shipowner Torvald Klaveness was listed as managing owner for a fleet of sixteen vessels totalling almost 350,000 deadweight tons (dwt). In part for legal reasons and partly for practical purposes, these vessels were owned by thirteen different companies.

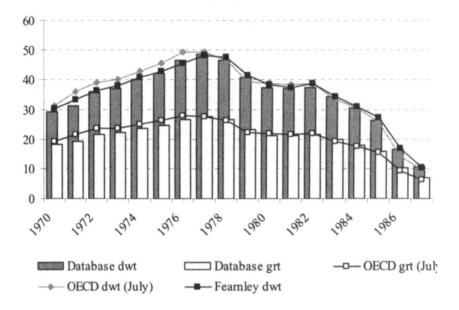

Figure 7.1: Norwegian Fleet, Various Sources, 1970-1987 (million grt and dead-weight tons [dwt])

Source: See text.

A related problem is the failure of the database – and the register on which it is based – to consider vessels that Norwegian companies chartered-in.[3] It would be impossible – from an information perspective – to take such tonnage into account. Moreover, the inclusion of chartered-in tonnage means that the concept of "the Norwegian fleet" would contravene common definitions. For instance, investment and employment related to chartered-in vessels cannot be compared with domestically-owned tonnage.[4] Thus, while the database gives a good depiction of Norwegian owners and their activities, it does not embrace all Norwegian companies involved in shipping.

---

[3]According to his biographers, John Fredriksen controlled a tanker fleet of four million dwt in the autumn of 1979, most of which had been chartered for one year; see Odd Harald Hauge and Gunnar Stavrum, *John Fredriksen – uautorisert biografi* [*John Fredriksen – An Unauthorised Biography*] (Oslo, 1991), 69. As a result, Fredriksen managed Norway's second largest fleet. But since the vessels were chartered rather than owned, Fredriksen's fleet is not included in the Veritas register or the database.

[4]As the chartered tonnage could be owned by other Norwegian shipping companies, including it in a measure of the fleet would lead to double-counting.

The degree of concentration in Norwegian shipping was low by international standards.[5] An analysis in the mid-1970s of seven European countries showed that all had more concentrated ownership structures than Norway. Among the Nordic countries, Denmark stood out because of the position of Rederiet A.P. Møller, which owned two-thirds of the Danish fleet. In Finland, the ten largest companies owned 85.3 percent of the fleet in 1976, and the corresponding figure for Sweden was 90.2 percent. The survey also included France, Germany and Belgium, where the ten largest companies' shares were 72.5, 56.5 and 94.6 percent, respectively. The authors of the study estimated that the ten largest Norwegian companies owned 46.6 percent of the fleet in mid-1973, about four percent higher than the database indicates.[6] Nevertheless, compared with other nations, ownership in the Norwegian merchant marine was spread among a large number of companies.

The shipping crisis, however, led to increased concentration. In 1970 there were 176 shipping companies with vessels of more than 5000 grt registered in Norway; by 1987 the number had fallen by more than two-thirds to only fifty-six; if we include firms with tonnage registered abroad, the figure increases to ninety-one. Whichever figure is used, it is clear that the decline was substantial. The reduction in the number of companies was particularly sharp before 1980 and in 1985-1986. There were four reasons for the decline. First, some companies merged – although the effect of this was minimal. Second, a number withdrew from shipping due to what they judged to be limited opportunities for profit. This factor was undoubtedly the most important. Third, some chose to register ships abroad; of these, some continued operations under foreign flags, while others gave up after a time. This development was reflected in the difference between the ninety-one companies that owned tonnage in 1987 and the fifty-six that owned Norwegian-registered tonnage. Finally, about twenty-five owners disappeared because they sold their vessels above 5000 grt. While some abandoned shipping altogether, others continued to operate smaller vessels.

The decrease in the number of shipowners was accompanied from the early 1980s by a reduction in average fleet size. This was partly due to the sale of a significant proportion of the largest tankers and dry bulk carriers. The decline was particularly evident after the Norwegian Guarantee Institute ended

---

[5]Ignacy Chrzanowski, Maciel Krzyzanowski and Krzysztof Luks, *Shipping Economics and Policy. A Socialist View* (London, 1979), 81-96.

[6]One reason for the discrepancy may be that the study underestimates the size of the Norwegian fleet; *ibid.*, 90-91. A comparison of the figures from table 3.10 of this study and those in Organisation for Economic Co-operation and Development (OECD), *Maritime Transport 1973* (Paris, 1974), indicates that the size of the aggregate Norwegian fleet used in the study is too low, leading to an overestimation of the degree of concentration.

its involvement. Figure 7.2 shows that the average size of the shipping companies in 1987 was not markedly larger than it had been in 1970.

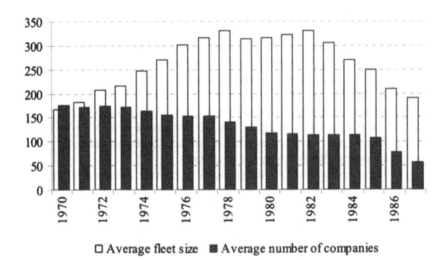

□ Average fleet size   ■ Average number of companies

Figure 7.2:   Number of Companies and Average Fleet Size, 1970-1987 ('000 dwt)

Note: Excludes companies that only owned drilling rigs.

Source: See text.

The fall in the number of shipowners led to increased concentration. The share of the fleet owned by Sig. Bergesen d.y. and Co. – the largest owner – grew steadily from 7.5 percent in 1970 to more than thirty-four percent in 1987.[7] The tendency towards increased concentration is evident from figure 7.3.

It might be fruitful to divide the period into two when analysing the structural changes. The shifts that occurred during the strong growth period prior to 1977 reveal transformations in the structure of Norwegian shipping in an expansive era, while subsequent developments illustrate the changes after the shipping crisis had left its mark.

---

[7]The twenty largest companies depicted in figure 7.3 are to the twenty biggest in any given year. Of the twenty largest companies in 1970, only eight were still included by 1987.

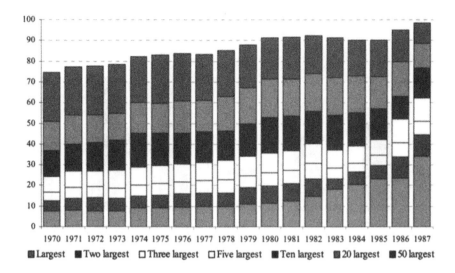

Figure 7.3: Concentration of the Fleet, 1970-1987 (dwt)

Source: See text.

Developments in the first period show that fleet growth was unevenly distributed among the various companies. Not surprisingly, the highest absolute growth occurred in the largest companies, while the strongest relative growth for the most part took place among the smaller companies. Nine of the ten largest companies in 1970 increased their fleets, most of them considerably.[8] Figure 7.4 shows that the ten largest companies increased their tonnage more than did the small and medium-size firms. Perhaps surprisingly, the development is more pronounced when grt rather than dwt is used. This can be explained by the fact that although the largest companies included shipowners such as Bergesen and Reksten, who invested solely in mammoth vessels, several of the larger companies, including Leif Höegh and Co. AS, Wilh. Wilhelmsen and Fearnley and Eger invested in general cargo tonnage, gas and vehicle carriers and drilling vessels, all of which were smaller. It was particularly the companies in the "11-50" bracket that invested in larger ships.[9]

---

[8]The exception was Fred. Olsen, who ranked ninth in 1970. He disposed of his tankers and large cargo ships, concentrating instead on passenger vessels and drilling rigs.

[9]The grt/dwt-conversion factor for tonnage growth in this category was almost 2.19, compared with approximately two for the other two groups depicted in figure 7.4. The conversion tended to be higher for larger than for smaller vessels.

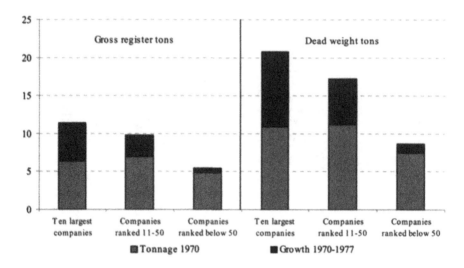

Figure 7.4: Tonnage Growth, 1970-1977 (million grt and dwt)

Source: See text.

The smallest category in figure 7.4 included some minor owners who soon left shipping and others who were rapidly expanding their fleets. Of the companies registered in 1970, fifty-two, with a total fleet size of 1.2 million grt, were no longer listed by 1977. Those which disappeared comprised about thirty percent of all shipowners in 1970 but owned less than seven percent of the tonnage. In 1970 the average fleet size of these companies was less that 25,000 grt, compared with slightly more than 100,000 grt for the entire fleet. The largest company to disappear was the Herness Shipping Company, which in 1970 ranked fifty-third with a five-ship fleet of 96,000 grt.

The reduction in the number of shipping companies was partly offset by new entrants. Of those registered in 1977, thirty-three were new since 1970 and three had acquired holdings larger than 100,000 grt.[10] In total, the new companies owned approximately one million grt of shipping, in addition to six offshore vessels. The aggregate fleet of the new entrants was slightly smaller than the fleets of the deleted companies in 1970. A closer examination of the new firms reveals that five were primarily offshore companies; three operated passenger vessels; three were linked to oil companies; and one was a bank that

---

[10]The number of companies registered in 1970 and 1977 were 176 and 158, respectively. The reason that the net reduction of nineteen (from fifty-two to thirty-three) companies diverges from the eighteen (176 to 158) found in the annual data is that one managing owner had two separate firms in the annual figures for 1977. These have been added in connection with the estimate of growth from 1970 to 1977.

had temporarily taken possession of a ship. Five of the companies were registered as shipowners in 1970 but did not own vessels larger than 5000 grt. And six of the companies had strong ties to extant firms in 1970.

The increased concentration in Norwegian shipping from 1970 to 1977 can be understood in terms of differences in the expansion rate of companies at various levels in the shipping hierarchy. The fleets of the fifty largest companies increased by approximately sixty percent over the period, compared with twelve percent for those ranked between fifty-one and 176 in 1970. Nine of the ten largest companies were particularly expansive, acquiring on average more than 500,000 grt each. Indeed, more than sixty-three percent of all expansion was accounted for by these nine firms (see table 7.1).

A discussion of growth in the period 1977-1987 is somewhat misplaced. Only eleven of the 158 companies registered in 1977 had more tonnage on Norwegian registry in 1987. The majority – 111 – had disposed of their Norwegian-registered fleets altogether. The share of the deleted companies thus increased from thirty percent in 1970-1977 to more than seventy percent from 1977 to 1987.[11] Since there were considerable changes in Norwegian shipping after 1977, a description of the industry is more difficult. The main reason is the turbulence, characterised by creditor involvements, bankruptcies, liquidations and reorganisations. One important question is how to treat firms that were severely affected by the crisis. Several went out of business but in the process laid the foundations for new firms. The following analysis focuses on the fleets rather than the legal companies of the managing owners.[12] For instance, the vessels registered in 1987 under Havtor Management have been compared with the 1977 fleet of the shipowner P. Meyer. The reorganisation of his company, based on the goodwill of the creditors and the Norwegian government, laid the foundation for the Havtor fleet. Another important aspect is the handling of companies with tonnage registered abroad. The liberalisation of the flag policy had important consequences for shipping. The results of an analysis of the structural changes in the Norwegian fleet differ substantially depending upon whether foreign-registered vessels are included.[13]

---

[11]The figures refer to comparisons of individual years. In aggregate, forty-five companies were deleted in the period 1970-1977, and 109 were lost during the years 1977-1987. The number of companies leaving the industry refers to those with tonnage registered in Norway.

[12]It was common for some of the companies to find themselves in economic difficulties, while others related to the same shipowner remained relatively unscathed. The degree to which the managing owner was answerable as a result of joint responsibility and reciprocity agreements varied.

[13]Due to the strict flag policy, it was largely unnecessary to make such a distinction before 1977.

Table 7.1
Companies with the Highest Absolute and Relative Growth, 1970-1977 (grt)

	Absolute				Relative		
Company	Rank 1970	Growth (grt)	Percent	Company	Rank 1970	Growth (grt)	Per-cent
S. Bergesen d.y.	1	1,240,886	105.4	Jan-Erik Dyvi	169	146,038	1536.6
H. Reksten	4	931,939	143.6	K.G. Jebsen	170	82,512	900.2
Anders Jahre	2	783,618	91.1	O.B. Sørensen	135	97,770	471.4
E. Rasmussen	10	586,540	163.9	A. Teigen	173	19,760	332.7
Wilh. Wilhelmsen	3	553,859	69.7	Gotaas-Larsen	146	45,703	293.6
H.Waage	22	530,061	244.3	J.P. Pedersen and Søn	94	133,433	287.1
Fearnley and Eger	6	464,128	78.1	H.Waage	22	530,061	244.3
L. Höegh	5	321,875	51.0	Skjelbreds Rederi	99	92,171	233.7
C.H. Sørensen	25	304,121	153.4	Simonsen and Astrup	133	50,034	233.1
O. Godager	38	294,559	227.7	O. Tøn-Nevold	129	50,927	230.9
Knut Knutsen.	15	276,718	108.2	O. Godager	38	294,559	227.7
Biørn Biørnstad and Co.	18	243,308	102.8	Mosvolds Rederier	81	126,884	219.1
Mosvold Shipping	8	195,523	43.7	A.H. Mathi-Esen	59	189,383	211.1
A.H. Mathi-Esen	59	189,383	211.1	Olaf Pedersen	157	24,578	199.3
Thor Dahl	7	177,408	35.2	Stove Shipping	82	108,154	187.2
J.M. Ugland	25	162,954	79.9	H.M. Wrangell	66	146,396	180.8
H.E. Hansen-Tangen	36	153,619	114.0	Haldor Virik	97	73,416	176.9
H.M. Wran-Gell	66	146,396	180.8	O. Schrøder	103	62,935	175.1
Jan-Erik Dyvi	169	146,038	1536.6	E. Rasmussen	10	586,540	163.9
Sigurd Herlofson	13	145,710	55.0	C.H. Sørensen	25	304,121	153.4

Note: Excludes companies formed between 1970 and 1977.

Source: See text.

Figure 7.5 shows how the reduction in the Norwegian fleet from 1977 to 1987 was distributed. This mirrors the previous period, although the reason for the increasing importance of the largest firms was that they experienced less severe contraction rather than greater expansion (see table 7.2). The difference was particularly evident for Norwegian-registered tonnage. The share owned by the ten largest companies in 1977 increased from approximately forty-three to almost sixty-one percent if only Norwegian-registered tonnage is considered, but when vessels registered abroad are included, it becomes less than half. [14] The difference shows that flagging-out was used more extensively by medium and small companies than by the shipping giants.

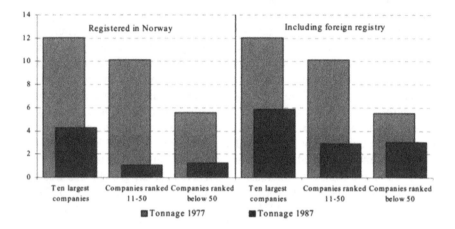

Figure 7.5: Changes in Tonnage, 1977-1987 (million grt)

Source: See text.

Table 7.2
Tonnage and Tonnage Decline, 1977-1987

	Norwegian Register			Norwegian and Foreign Registers	
	1977	1987	Decline %	1987	Decline %
Ten Largest Companies	12,018,309	4,295,422	64.3	5,880,759	51.1
Companies ranked 11-50	10,153,007	1,046,582	89.7	2,932,410	71.1
Companies ranked below 50	5,564,929	1,272,733	77.1	3.027,381	45.6

Source: See text.

A closer examination of the thirty-three new entrants in the 1977 figures reveals the shifting fortunes of the industry. More than two-thirds had disappeared by 1987, but five continued to expand when others were disinvesting. Only eleven increased their Norwegian-registered tonnage from 1977 to 1987. One of them, Einar Rasmussen, was among the country's ten largest

---

[14] These figures deviate from those in figure 7.3 because they refer to the ten largest companies in 1977, whereas figure 7.4 shows the share of the fleet owned by the ten largest companies in any given year.

shipowners in 1977, but his tonnage increase was relatively small.[15] The expanding companies also included the cruise operators Anders Wilhelmsen and Co. and Lauritz Kloster/Kloster Cruise AS, ranked fifty-ninth and ninety-ninth, respectively, in 1977. None of the other expanding firms was among the 100 largest in 1977. The eleven that grew experienced only modest growth totalling approximately 395,000 grt, an aggregate increase of merely twenty-three percent.[16] Two of them, both among the newcomers in 1977, operated only drilling vessels. Even when foreign-registered vessels are included, the contrast between the periods is striking. In the first, seventy-eight firms increased their tonnage, forty-four reduced their fleets and fifty-two left shiping. This represented forty-four, twenty-five and thirty percent of the companies, respectively.[17] From 1977 to 1987 only nineteen firms increased their tonnage; forty-three scaled down their fleets; and ninety-two abandoned shipping or focused on ships smaller than 5000 grt. Only twelve percent increased their fleets, while twenty-eight percent reduced their tonnage. The largest category – sixty percent – consisted of companies deleted from the registry.

For the companies increasing their tonnage, the growth was less pronounced than in the previous period (see table 7.3). From 1970 to 1977 the expanding companies increased their fleets by almost eleven million grt. From 1977 to 1987 the aggregate growth of those that increased their fleets was 1.6 million grt, which would only have represented about three-quarters of the growth of the two most expansive companies in the previous period. Only four companies increased their fleets to such an extent that they would have been included among the thirty firms with the highest absolute growth in the period 1970-1977. In short, from 1977 to 1987 companies that expanded their fleets were in a minority. In general, growth was more modest and less widespread than in the earlier period. Fittingly, a ferry company's investment in a larger vessel warranted its inclusion among the fastest-growing Norwegian shipowners.

A comparison of the expanding companies in the periods 1970-1977 and 1977-1987 reveals three interesting aspects. First, only three had been among Norway's fifty largest shipowners in 1977; instead, growth clearly occurred at the lower levels in the shipping hierarchy. Second, four which expanded their fleets from 1977 to 1987 were among those with the highest growth in the previous period as well. Leif Höegh and Co. AS and J.M. Ug-

---

[15]The growth of 1.8 percent refers to the combined tonnage of Einar Rasmussen and Rasmussen Management in 1987. In the international figures, which exclude drilling vessels, there was a slight reduction in the Rasmussen fleet.

[16]If the largest company, Einar Rasmussen/Rasmussen Management, is excluded, average tonnage growth increases to 113 percent.

[17]Two companies maintained the *status quo* from 1970 to 1977.

land were among the firms with the largest absolute tonnage increases from 1970 to 1977, whereas Kristian Gerhard Jebsens Skipsrederi AS and Ole Schrøder and Co. AS had high relative growth. Third, Fred. Olsen, who recorded the largest absolute fleet reduction from 1970 to 1977, was among the fastest growing companies in the next period. From 1970 to 1977 his fleet declined by approximately 270,000 grt as he left the tanker sector, but his fleet increased by about 260,000 grt in the following decade as he re-entered the sector with three Liberia-registered tankers.

Table 7.3
Companies with Expanding Fleets, 1977-1987

Company	Absolute Rank	Growth (grt)	Company	Relative Rank	Growth %
K.G. Jebsen	73	279,468	Lauritz Kloster	99	368.1
Fred. Olsen and Co.	75	261,116	K.G. Jebsen	73	304.8
Ugland Management	88	183,047	Fred. Olsen and Co.	75	300.7
Lauritz Kloster	99	168,629	Chr. J. Reim	124	296.3
AS Bill	102	94,978	Ugland Management	88	286.0
Anders Wilhelmsen	59	94,186	AS Bill	102	223.5
J.M. Ugland	17	83,582	Gerrards Rederi AS	120	218.2
Kristian Jebsen	66	76,491	Peter Y. Berg	154	215.2
Ole Schrøder	70	71,170	Larviks-Frederikshavnferjen	150	120.4
Leif Höegh and Co.	6	70,282	Ugland Shipping.	103	96.9
Chr. J. Reim	124	55,531	Anders Wilhelmsen	59	76.4
Torvald Klaveness	28	54,937	Iver Bugge	113	75.2
Gerrards Rederi	120	44,689	Ole Schrøder	70	72.0
Ugland Shipping	103	39,843	Kristian Jebsen	66	70.4
Iver Bugge	113	21,096	William Hansen	129	29.6
Peter Y. Berg	154	10,915	Torvald Klaveness	28	22.9
Larviks-Frederikshavnferjen	150	8189	J.M. Ugland	17	22.8
William Hansen	129	4864	Leif Höegh and Co.	6	7.4
Christian Haaland	54	2502	Christian Haaland	54	1.9

Note: The fleet assigned to J.M. Ugland in 1987 comprises Uglands Rederi AS; AS Ugland Shuttle Tankers; and Ugland Construction Co. AS. Ugland Management Co. includes this company as well as Andreas Ugland Autoliners AS and Andreas Ugland Ventures AS. The fleet assigned to K.G. Jebsen includes the fleets of this company and Gearbulk Ltd., whereas the fleet of Kristian Jebsen refers to AS Jebsens Ship Management. Ole Schrøder refers to Osco Shipping, and Lauritz Kloster to Kloster Cruise AS. In determining the 1977 rank, companies operating drilling vessels only are excluded, bringing the total number to 154 rather than 158.

Source: See text.

Table 7.4
Comparison of the Growth of Shipping Companies

Company Size	Increase 1970-1977 Percent (grt)	Decline 1977-1987, Norwegian Register	Decline 1977-1987, All ships
Ten largest companies	78.2	-64.3	-51.1
Companies ranked 11-50	40.2	-89.7	-71.1
Companies ranked below 50	12.1	-77.1	-45.2

Note: The figure for the reduction of the foreign-registered fleet for companies
           ranked lower than fiftieth differs from the one in table 7.2 by 0.4 percentage
           points. The reason is that companies only operating drilling vessels were in-
           cluded in the 1977 fleet in table 7.2 but are excluded in here.

Source: See text.

The contrast between the expansive and contractionary periods is
striking. In the former, large and medium-sized companies grew most rapidly,
with smaller firms losing ground. From 1977 to 1987, however, it was mainly
the largest companies, and some smaller ones, that maintained their relative
position (see table 7.4). The fleets of medium-sized firms underwent the most
severe contraction. Moreover, the low aggregate growth rate of the companies
ranked below fiftieth from 1970 to 1977 was partly affected by the demise of a
considerable share of those registered – all of the fifty-two companies deleted
in the 1977 figures were in this category. From 1977 to 1987 the other categories also contained disappearing companies.

Two of the ten largest shipowners in 1977, with combined fleets of
1.4 million grt, had left shipping by 1987. In terms of tonnage, the bulk of the
erosion occurred among those ranked between eleventh and fiftieth. Fewer
than half of these forty firms remained by 1987. Over the ten-year period,
twenty-three companies, with an aggregate tonnage of more than 5.5 million
grt, were deleted from the registry. Of the 104 firms below fiftieth, sixty-
seven, with an aggregate tonnage of less than 3.2 million grt in 1977, were
deleted. The average size of the companies disappearing was considerably larger than the 23,400 grt recorded in the period 1970-1977. If we include foreign-registered tonnage, the average size of the deleted firms was approximately 110,000 grt. If we look at those deleted from the Norwegian registry,
the average size was 121,500 grt. Moreover, companies of various sizes, not
only those below 100,000 grt, departed.

The ninety-two deleted companies comprised approximately sixty percent of all firms in 1977 but owned less than thirty-seven percent of the tonnage. Those that had no tonnage left on Norwegian registry constituted more
than seventy percent of all companies in 1977 and owned approximately half
the fleet. A comparison with the figures for the period 1970-1977, which were
thirty percent of the companies and less than seven percent of the tonnage,

shows that a larger portion of the firms disappeared. Moreover, they owned a considerably larger share of the Norwegian fleet.

Table 7.5
Companies Disappearing by Size, 1977-1987

Size (grt)	No Tonnage in Norway	Tonnage 1977	No Tonnage in Norway	Tonnage 1987
More than 500,000	5	4,044,542	3	1,961,452
200,000-500,000	12	3,409,846	11	3,186,703
100,000-200,000	25	3,592,681	21	3,046,582
50,000-100,000	21	1,500,843	16	1,114,728
Less than 50,000	48	947,088	41	820,979
Total	111	13,495,000	92	10,130,444

Source: See text.

Despite the severe contraction of the Norwegian fleet, some new companies entered shipping during the depression. In 1987, thirteen firms that had not been registered in 1977 owned ships registered in Norway.[18] Four were offshore companies, operating mainly drilling vessels. The aggregate tonnage of the other nine amounted to some 500,000 grt. If we include those managing vessels registered abroad, the figure increases to seventeen companies with an aggregate tonnage of approximately 1.26 million grt.[19] Of the new entrants, Skibs-AS Karlander and O.H. Meling had been registered in the early 1970s, but had temporarily fallen below the 5000 grt limit. Two of the firms had previously operated smaller ships but acquired vessels above 5000 grt between 1977 and 1987. Two of the new owners were manufacturers. The remaining eleven new entrants had close links to other shipping companies, such as J.O. Odfjell AS, which was established in connection with the division of AS Rederiet Odfjell between two branches of the Odfjell family.

The period 1977-1987 was extremely turbulent, with a strong reduction in both tonnage and the number of companies. The Norwegian-registered fleet, measured in dwt, declined by more than three-quarters between 1977 and 1987, while the number of companies with ships registered in Norway fell from 154 to fifty-six. This was partly mitigated by the registration of vessels abroad, but the Norwegian fleet still fell by more than half, although the reduction was not evenly distributed across the various regions of the country.

---

[18]A dynamic, as opposed to a cross-sectional, analysis shows that ninety-four new companies were registered in the period 1978-1987, but the vast majority disappeared by 1987.

[19]The offshore companies have been excluded from these figures.

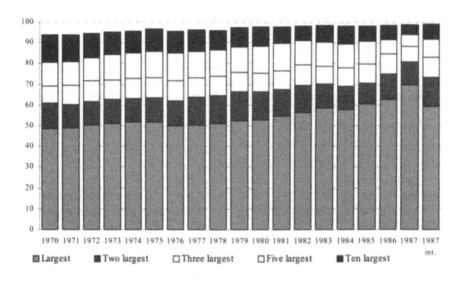

Figure 7.6: Concentration in Norwegian Shipping Ports, 1970-1987

Note: "1987 int." includes foreign-registered vessels managed by Norwegian
                companies which have been assigned to the port where the firm was
                based. Two changes have been made relative to Veritas and the list in ap-
                pendix 3, where Sig. Bergesen d.y. and Wilh. Wilhelmsen had their
                home ports listed as Stavanger and Tønsberg, respectively. Here they are
                included in Oslo where their main offices had been for decades. This
                categorisation will be used in the rest of the analysis.

Source: See text.

     Spatial dispersion of the ownership of the fleet has long been a charac-
teristic of the Norwegian shipping industry. Shipping has been important in
ports all along the country's extensive coastline, and the widespread pattern of
shipping can be explained by the manner in which the industry evolved. The
location of shipping activities was based initially on access to raw materials for
shipbuilding and a stable labour supply. This changed during the nineteenth
century, as access to capital markets, proximity to communications and trading
centres, organisational competence and dependence upon a qualified domestic
labour force became more important.[20] Nonetheless, shipping companies con-

---

    [20]See Helge W. Nordvik "The Shipping Industries of the Scandinavian Coun-
tries: 1850-1913," in Lewis R. Fischer and Gerald E. Panting (eds.), *Change and Adaptation in Maritime History: The North Atlantic Fleets in the Nineteenth Century* (St.
John's, 1985), 117-148.

tinued to be situated in a large number of ports. The shipping crisis to some extent changed this feature, as the importance of the largest ports increased, whereas activity was discontinued in several smaller ports.[21] Moreover, the database shows that some regions were harder hit by the crisis than others.

The number of ports in which vessels above 5000 grt were registered almost halved from 1970 to 1987, falling from twenty-seven to fourteen. The ports disappearing during the 1970s were mostly insignificant, comprising together less than two percent of the fleet at the start of the decade. The ports deleted in the 1980s, however, were more important. Arendal. For example, was among Norway's five largest in the 1970s, owned almost five percent of the fleet in 1984, but had no tonnage above 5000 grt on Norwegian registry by 1987. At the same time, the importance of the largest ports increased. Figure 7.6 shows the increased concentration in Norwegian shipping by port. This reflected not only a reduction in the number of ports – the number outside the "ten largest" bracket fell from seventeen in 1970 to four by 1987 – but also the growing importance of the largest shipping companies. Indeed, the tonnage of the largest firm, Sig. Bergesen d,y., would have been Norway's second largest "port" in 1987, two and one-half times larger than third-place Kristiansand.

The regional changes are evident in figure 7.7. In 1970 about half the fleet was owned by firms based in Oslo, while Oslofjord, the South Coast and Western Norway each had between fourteen and nineteen percent. The picture was similar in 1980, although Oslo's share had increased by about four percentage, as had Oslofjord. But the 1980s were marked by more prominent shifts. Oslofjord's tonnage continued to decline, and its share of the fleet fell to about four percent by 1987, in part because the number of companies fell from twenty-three to eight. Moreover, average tonnage fell from 111,500 grt in 1970 to less than 40,000 grt in 1987, after peaking at 184,000 grt in 1978. The share owned by firms in Western Norway was constant until 1980 but plunged from almost sixteen to about eleven percent from 1984 to 1985. This was due to a reduction in the Bergen fleet of more than 500,000 grt (twenty-eight percent) and a similar decline in Haugesund (which fell by almost three-quarters). In the latter, the decline was due largely to the sale of several large tankers, whereas the reduction in the former was the result of a combination of the scrapping or sale of large tankers and the transfer of tonnage to foreign flags.

---

[21]See appendix 3 for a detailed list of the Norwegian fleet in various ports. The trend towards concentration is evident even though the 5000-grt limit was relatively more strict in 1970 than in the mid-1980s due to the increase in average vessel size.

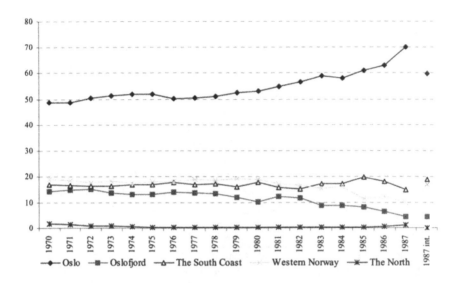

Figure 7.7: Share of the Fleet by Region, 1970-1987 (percent)

Note: Oslofjord includes Drammen, Fredrikstad, Horten, Larvik, Moss, Pors-
            grunn, Sandefjord, Skien and Tønsberg; South Coast includes Arendal,
            Farsund, Flekkefjord, Grimstad, Kristiansand, Lillesand, Mandal and
            Tvedestrand; Western Norway includes Bergen, Egersund, Haugesund,
            Skudeneshavn and Stavanger; and North includes Bodø, Kirkenes, Narvik,
            Stokmarknes, Tromsø, Trondheim and Ålesund.

Source: See text.

        A comparison of the Norwegian-based and Norwegian-owned fleets in
1987 reveals some interesting features (see table 7.6). First, the share of Oslo
is reduced when foreign-registered vessels are taken into account, whereas the
shares of the South Coast and Western Norway increase. This is partly because
shipowners in the latter regions were more eager than those in Oslo to register
tonnage abroad. The share of tonnage owned by shipowners in Grimstad is
more than doubled when we look at Norwegian-owned rather than Norwegian-
registered tonnage. Similarly, the share of the fleet based in Bergen increases
from approximately eight to nearly fourteen percent. Second, when tonnage
registered abroad is included, there is no significant reduction in the share of
the Norwegian fleet based on the South Coast or Western Norway compared
with 1970. Instead, there is a small increase from 16.8 to 18.9 percent for the
South Coast, and a corresponding decline from 18.8 to 16.9 percent for Western Norway.

Table 7.6
Tonnage in 1987 as a Share of 1970 Fleet and Peak Fleet by Region

	1987/1970	1987/peak	1987 int./1970	1987 int./peak
Oslo	56.1	35.6 (1977)	88.2	56.0 (1977)
Oslofjord	12.4	8.4 (1977)	21.9	14.9 (1977)
South Coast	34.4	22.0 (1976)	80.8	51.8 (1976)
Western Norway	20.6	13.5 (1977)	64.8	42.6 (1977)
North	24.6	73.7 (1978)	6.1	18.2 (1978)
Aggregate	39.1	25.7 (1977)	71.9	47.2 (1977)

Source: See text.

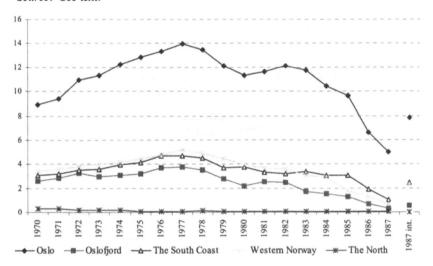

Figure 7.8: Tonnage in Various Regions, 1970-1987 (million grt)

Source: See text.

The tonnage of Oslo, Oslofjord and Western Norway peaked in 1977, just as did the aggregate Norwegian fleet. The tonnage registered on the South Coast peaked the previous year, and the North's fleet reached its apex in 1970. Figure 7.8 also shows that the temporary cessation of the decline of the Norwegian fleet in the early 1980s was largely a result of the growth in the Oslo and Oslofjord fleets, and occurred during the continuing reduction of the tonnage registered on the South Coast and in Western Norway. The relative reduction of the fleets reveals large regional differences. The reduction of the fleet registered in the North occurred earlier and was more pronounced than for the rest of the country; from 1970 to 1976, tonnage fell by more than ninety percent. This partly reflected the relatively small number of vessels registered in the various ports, which implies that the disposal of only a few

vessels might have large effects. The reduction was influenced by the exit of
ports such as Bodø, Kirkenes, Stokmarknes and Ålesund, as well as a decrease
in tonnage registered in Tromsø. Trondheim was the only port in the North
where the tonnage was relatively stable, but its share of the Norwegian fleet
fell as a result of the considerable expansion elsewhere in the country. Apart
from the relatively insignificant fleet in the North, the highest relative decrease
was recorded in Oslofjord, where 1987 tonnage amounted to merely 12.4 percent of that registered in 1970 and only 8.4 percent as much as ten years previously. The situation was almost as severe if we adjust for tonnage registered
abroad. If we take such vessels into account, the situation in 1987 was not as
grave for the three main regions, Oslo, the South Coast and Western Norway.
While none had fleets as large as in 1970, let alone as in the peak in the mid-
1970s, the decrease was less than twenty percent in Oslo and on the South
Coast and approximately one-third in the case of Western Norway.

The heterogeneity of Norwegian shipping becomes evident when data
on the number of companies are seen in relation to the regional tonnage distribution. Figure 7.9 reveals that there were considerable differences between the
"average companies" in the various regions. The North was distinguished not
only by a relatively low number of companies but also by firms that were
smaller than the national average.

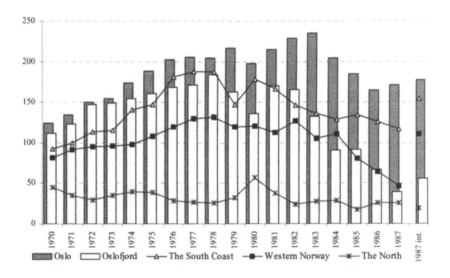

Figure 7.9: Average Size of Companies by Region, 1970-1987 ('000 grt)

Source: See text.

Table 7.7
Largest Ports, Percent of Norwegian Fleet, 1970-1987 (grt)

	-70	-71	-72	-73	-74	-75	-76	-77	-78	-79	-80	-81	-82	-83	-84	-85	-86	-87	-87 int.
Oslo	48.6	48.7	50.3	51.2	51.9	52.0	50.1	50.4	50.9	52.6	53.1	54.8	56.5	58.7	57.9	60.8	62.8	69.8	59.7
Bergen	12.2	11.7	11.4	11.6	11.2	11.5	12.2	13.4	13.6	13.9	13.6	13.0	12.8	11.5	11.3	9.3	9.6	7.6	13.9
Sande-fjord	8.4	9.1	9.2	9.3	9.5	9.8	9.3	9.3	9.1	8.7	8.9	8.9	9.9	7.0	6.4	6.0	4.9	2.9	2.4
Kristian-sand	8.0	7.8	7.7	8.2	8.0	8.1	9.0	9.0	9.3	9.1	8.8	8.9	7.6	8.5	8.7	9.8	12.1	11.0	9.5
Hauge-sund	3.4	3.7	3.2	3.6	3.6	3.5	3.6	3.5	3.4	3.6	3.5	2.9	2.8	2.7	3.4	1.0	1.1	1.6	1.1
Arendal	3.2	3.1	3.4	3.2	4.2	4.3	4.4	3.4	3.3	3.6	4.1	4.1	4.4	4.7	4.9	4.8	1.9		1.7
Grim-stad	2.7	2.7	2.6	2.3	2.4	2.5	2.2	2.3	2.4	1.8	1.8	1.9	2.1	2.5	2.1	2.3	2.3	2.9	6.2

Source: See text.

Table 7.8
Largest Ports by Rank, 1970-1987

	-70	-71	-72	-73	-74	-75	-76	-77	-78	-79	-80	-81	-82	-83	-84	-85	-86	-87	-87 int.
Oslo	1	1	1	1	1	1	1	1	1	1	1	1	1	1	1	1	1	1	1
Bergen	2	2	2	2	2	2	2	2	2	2	2	2	2	2	2	2	3	3	2
Sande-fjord	3	3	3	3	3	3	3	3	3	3	3	3	3	4	4	4	4	4	5
Kristian-sand	4	4	4	4	4	4	4	4	4	4	4	4	4	3	3	2	2	2	3
Hauge-sund	5	5		5				5	5	5									
Arendal			5		5	5	5				5	5	5	5	5	5			4
Grimstad																	5	5	5

Source: See text.

The combination of tonnage figures and data on the number of companies sheds light upon several characteristics. First, the demise of the Oslofjord fleet can be understood in terms of the dramatic reduction in the average size of the companies, which fell by more than seventy-five percent from 1981 to 1987; this was more important than the effect of the decrease in the number of firms from fifteen to eight. Second, the overall effect of the large reduction of companies on the South Coast was partly offset by the fact that their average size actually increased between 1970 and 1987. Third, the large difference in average size between the firms in Oslo and those in the North indicates that there were significant structural differences even within the domestic industry. Moreover, the average size of the companies in all of the three major shipping regions increased from 1970 to 1987 when foreign-registered vessels are included.

The regional changes can be analysed in even more detail on the port level. This shows that the regional categorisation disguises variations among ports within the same region. Perhaps the best example is the South Coast, where the decline of the fleet based in Arendal was basically neutralised by increases in the Grimstad and Kristiansand fleets. Annual figures reveal that seven different ports were at some point among the five largest between 1970 and 1987. Increased concentration implies that the importance of these ports grew during the period. The largest growth occurred in Oslo, which increased its share of the Norwegian-registered fleet from 48.6 percent in 1970 to almost seventy percent in 1987. The relative growth was of nearly the same magnitude in Kristiansand, while Sandefjord suffered the largest decline.[22]

Table 7.7 confirms the trend towards concentration in Oslo. Moreover, the latter two columns show the differing impact when foreign-registered vessels are included in the 1987 figures. In the case of Bergen, the share of the fleet almost doubles when vessels registered abroad are taken into account. This can be contrasted with decreases in Haugesund, Kristiansand and Sandefjord. The difference suggests that shipowners in some cities were more eager to take advantage of the liberalisation of Norwegian flag policy. Accordingly, the reduction of the fleet in places such as Bergen during the first half of the 1980s will be less dramatic if foreign-registered vessels are considered.

Table 7.8 shows that the relationship among the three largest ports was stable until 1983, when Kristiansand seized Sandefjord's position as the third largest port, before besting Bergen two years later. Sandefjord's share of the fleet fell dramatically after 1982 and by 1987 was reduced by more than

---

[22]The situation in Arendal was even more grave when only Norwegian-registered tonnage is included. By 1987, no tonnage above 5000 grt was registered there. This was partly because Arendal shipowners had taken advantage of access to foreign registries: when foreign-registered vessels are included, Arendal's share of the Norwegian fleet was approximately halved relative to 1970, whereas Sandefjord's share fell by more than seventy percent.

two-thirds. The position as fifth largest port alternated between Haugesund and Arendal until 1980, when the latter consolidated its position until it was surpassed by Grimstad in 1986. The table also reveals that the rank of the various ports depended upon whether or not foreign-registered vessels are included.

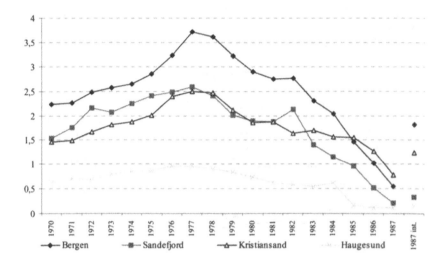

Figure 7.10: Tonnage in the Most Important Ports, 1970-1987 (million grt)

Note: For the development of Oslo, see the regional analysis.

Source: See text.

          The data in figure 7.10 and table 7.9 confirm the trend in the aggregate figures and in connection with the regional development. Again, there are large differences between the various ports with regard to the extent to which their fleets declined. When foreign-registered tonnage is taken into account, the decline was less than twenty percent for Oslo, Bergen and Kristiansand. The fact that their decline was less than the average for the Norwegian fleet reflects the increased concentration. Moreover, some ports, notably Grimstad, actually increased their tonnage relative to the situation before the crisis.
          Although the Norwegian fleet declined substantially, particularly compared with the peak, the situation was not as grave as the figures suggest. Some reduction could have been expected because the world fleet declined from 1982 onwards.[23] Another mitigating factor is that a ton of shipping in 1970 was different from a ton of shipping in 1987. This is because the Norwe-

---

          [23]The reduction of the world fleet, however, was smaller than for any of the Norwegian regions.

gian fleet in the late 1980s consisted of far more specialised vessels than in 1970. Specialised ships are more costly to build but have a correspondingly higher revenue potential. This implies, on the one hand, that the value of the fleet – or rather that the value of each unit of tonnage – was higher in the mid-1980s than in the early 1970s. Furthermore, several owners had diversified their operations from the bulk sector. Norwegian owners were leading within market segments such as supply vessels, large gas tankers and roll on-roll off ("ro-ro") carriers, as well as in chemical, cruise and open-hatch bulk shipping.

Table 7.9
Tonnage in 1987 as Share of the 1970 and Peak Fleet by Port

	1987/1970	1987/peak	1987 int./1970	1987 int./peak
Oslo	56.1	35.6 (1977)	88.2	56.0 (1977)
Bergen	24.2	14.5 (1977)	81.6	49.0 (1977)
Sandefjord	13.5	8.0 (1977)	20.8	12.4 (1977)
Kristiansand	53.6	31.4 (1977)	85.0	49.8 (1977)
Haugesund	18.4	11.8 (1977)	24.1	15.6 (1977)
Arendal			39.3	19.2 (1976)
Grimstad	41.1	31.9 (1977)	163.3	126.8 (1977)
Norway	39.1	25.7 (1977)	71.9	47.2 (1977)

Source: See text.

The first part of the period 1970-1987 was characterised by strong expansion, while the second part was marked by a sharp reduction in the number of companies and tonnage on Norwegian registry. From 1970 to 1987 the number of firms with ships registered in Norway fell by more than two-thirds, while the mass departure of Norwegian-registered vessels sliced the size of the fleet by more than three-quarters relative to the peak year. But a focus on the number of companies and aggregate tonnage may paint an overly negative picture. On the one hand, developments after 1987 proved the viability of Norwegian shipowners once the introduction of the Norwegian International Ship Register reversed some of the slippage. On the other, Norwegian shipping to some extent had undergone a positive transformation: large tankers and bulk carriers, for which the market had been disastrous, played a less important role as owners focussed on some of the more advanced segments of the industry.

The previous analysis used grt, and to a lesser extent dwt, to depict changes in the Norwegian fleet. But the use of tonnage cloaks some of the sophistication – or lack thereof – of the vessels. In shipping, as in most other major industries, there is a strong correlation between sophistication and the quality, price and value of the factors of production. The large units that dominate the tanker and bulk carrier sectors are comparatively inexpensive. For example, in 1970 a 30,000-dwt product tanker was as expensive as a bulk carrier twice this size because of the former's more sophisticated technology.

Whereas the cost per dwt for the tanker would have been approximately US $370, the bulk carrier would have cost about US $175 per dwt – and the cost per dwt of a large tanker would have been even lower.[24] The relative new building price of the various types of vessels fluctuated with conditions in various market segments.[25] Nevertheless, although a random year might give a faulty impression of price, the pattern was clear: cost per ton varied considerably depending upon the type of vessel ordered. The pattern behind the price differences was simple. Smaller vessels were more expensive per ton than larger ones, and specialised tonnage was more expensive than standard.

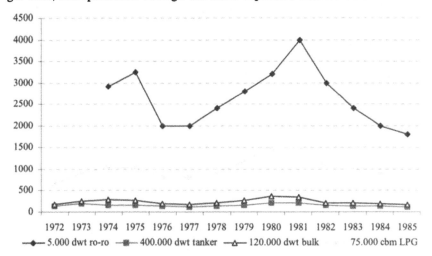

Figure 7.11: Differences in the Price of New Builds, 1972-1985 (US$ per dwt)

Note: The price of a 5000 dwt "ro-ro" vessel is only available from 1974 onwards.

Source: Fearnley and Eger, *Review*, various years, table headed "Contracting Prices for New Buildings."

Figure 7.11 reveals the large variations in cost per ton for various vessel types and sizes. The same pattern was evident in the second-hand market, although the picture was not quite as distinct. This is because the value of second-hand tonnage is more responsive to changes in the freight market, and the inclusion of age, and in particular condition, shows larger variations be-

---

[24]The figures are from Fearnley and Egers Chartering Co. Ltd., *Review 1972* (Oslo, 1973), 16.

[25]Demand factors play some role as well because the potential supply of the various types of vessels is limited.

tween vessels of otherwise identical types and sizes. The difference in the price and value of the various types of tonnage indicates that aggregate measures based on grt or dwt may distort actual developments. Indeed, when fleet structure is considered, the underlying figures give a better indication of both the type of shipping undertaken and the value of the vessels.

The structure of the fleet was transformed in two important ways. The first was a continuation of the long-term pattern since World War II: large tankers, combination carriers and bulk vessels became increasingly important. This trend continued until the final vessels ordered during the boom in 1973 were delivered around 1977/1978. The second commenced at the start of the 1980s. As the fleet declined, most tonnage disposed of comprised large tankers, combination carriers and bulk vessels. As well, Norwegian shipowners channelled much of their investment into more specialised types of tonnage, including gas tankers, chemical tankers and passenger vessels.

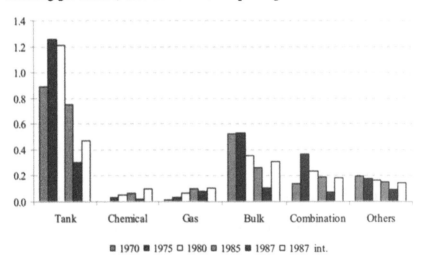

Figure 7.12: Tonnage in the Norwegian fleet, Selected Years, 1970-1987 (million grt)

Note: The figures for 1987 refer to the Norwegian-owned fleet. The reduction of tanker tonnage mentioned in the text refers to the Norwegian-registered fleet. The figures for "other vessels" include passenger, general cargo and other vessels that do not fit into the main categories.

Source: See text.

Figure 7.12 illustrates the structural changes in the Norwegian fleet. Tanker and combination carrier tonnage increased until 1975 but declined considerably thereafter. In the early 1970s, tankers, combination carriers, bulk carriers and general cargo vessels comprised more than ninety-eight percent of the Norwegian fleet. Seventeen years later this had fallen to approximately eighty percent, largely as a result of the growing proportion of gas carriers, chemical tankers and passenger vessels. The amount of tanker tonnage on the Norwegian register in 1987 was more than three-quarters below what it had been in 1975. The figures are even more conspicuous if the peak in 1977 is used as the basis for comparison. The most pronounced growth occurred in chemical tankers, gas tankers and passenger vessels. When the development portrayed in figure 7.12 is considered in terms of prices and values, it is evident that the fall in the value of the Norwegian fleet per grt must have been less than the aggregate figures indicate. This can be inferred by the increasing importance after 1975 of vessels that were relatively expensive per ton.

Although there is no universal measure of the value of various types of shipping tonnage, there have been attempts to examine this in terms of shipbuilding costs. An OECD working party developed a measure to establish comparable figures for the level of activity in shipbuilding based upon the grt but including a coefficient that adjusted for the demand for labour relative to a given "standard vessel." Using a relatively detailed system of these conversion coefficients, compensated gross register tonnage (cgrt) takes into account some of the price differentials. Both the type and size of the vessel affect the calculations. Although the method is intended to show structural traits in shipbuilding, it may be applied to the shipping sector as well.

The following analysis is based on the assumption that cgrt can be used to evaluate changes in the Norwegian merchant marine. Although this measure is relatively crude, it gives a better picture of the most important structural changes than analyses based only on grt or dwt. A conversion of the data on the Norwegian fleet from grt to cgrt has two important implications. First, the importance of large tankers, bulk carriers and combination vessels decreases because their coefficients are fairly low. Second, the decline of the Norwegian fleet becomes considerably less obvious, as vessels with low conversion factors were replaced largely by those with higher coefficients.

Table 7.10 shows the conversion factors for the various types of vessels and sizes. The large effects of a conversion may be illustrated by the extreme examples in the case of a fleet of one million grt. If this had consisted solely of mammoth tankers, its compensated tonnage would be 300,000 cgrt. On the other hand, if the fleet was comprised solely of specialised vessels like tugs, offshore supply vessels and drilling ships, its compensated tonnage would be five million cgrt. Indeed, the compensated size of the fleet would be doubled if it consisted of product carriers in the 30,000-50,000 dwt range rather than crude carriers above 250,000 dwt.

A conversion of the tonnage in the Norwegian fleet yields dramatically different results from a straightforward addition of the grt or dwt of the various vessel groups. Figure 7.13 is based upon the same data as figure 7.12 except that the vessels have been assigned to their various size classes, measured by dwt, and the compensated gross tonnage has been calculated by means of the relevant grt figures. The most conspicuous difference is the increased importance of general cargo vessels, passenger ships, ferries and specialised tonnage. By 1985, this group had surpassed tankers, despite the fact that the tankers were three times as important in a comparison based upon grt.

Table 7.10
Conversion Factors from OECD, Late 1970s (dwt)

Vessel Type/ Size	CF	Vessel Type/ Size	CF	Vessel Type/ Size	CF	Vessel Type/ Size	CF
Crude Oil Tankers		**Bulk Carriers**		**Reefers**		**LNG**	
10-30,000	0.65	10-30,000	0.60	4-10,000	2.00	4-10,000	1.60
30-50,000	0.50	30-50,000	0.50	10,000+	1.40	10-30,000	0.90
50-80,000	0.45	50-100,000	0.45	**Small Tankers/ Bulk Carriers**		30-50,000	0.70
80-160,000	0.40	100,000+	0.40	4-10,000	1.80	50,000+	0.50
160-250,000	0.35	**Combination Carriers**		**Full-container/ fastliners**		**Ro-ro/Car carriers**	
250,000+	0.30	10-30,000	0.65	4-10,000	1.40	4-10,000	2.00
Product Carriers		30-50,000	0.55	10-30,000	0.90	10,000+	1.60
10-30,000	0.80	50-100,000	0.50	30,000+	0.80	**Ferries/pass.**	
30-50,000	0.60	100,000+	0.40	**LPG and chemical**		All sizes	2.50
50,000+	0.50	**General Cargo**		4-10,000	1.60	**Other non-cargo**	
		4-10,000	2.00	10-30,000	1.00	All sizes	5.00
		10,000+	1.40	50,000+	0.80		

Note: "CF" refers to the conversion factors.

Source: Norges Offentlige Utredninger (1978:13), *Skipsfartsnæringen* [*The Shipping Sector*], 17-18.

The use of compensated measures changes the overall development of the fleet dramatically. Measured in grt, the Norwegian-registered fleet fell by 62.2 percent between 1970 and 1987 and by more than seventy-two percent from 1975 to 1987. Measured in cgrt, the declines are reduced to fifty-five and 62.5 percent, respectively. The conventional figures, by disregarding structural changes, overstate the decline, while the compensated measures show that the reduction was less pronounced than previously believed. Measured in cgrt, and including Norwegian tonnage registered abroad, the fleet was reduced by 27.6 percent relative to 1975, but by only 13.4 percent from 1970 to 1987.

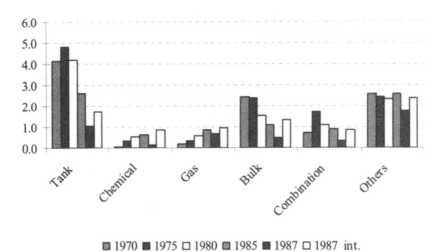

□ 1970 ■ 1975 □ 1980 ▨ 1985 ■ 1987 □ 1987 int.

Figure 7.13: Tonnage in the Norwegian Fleet, Selected Years, 1970-1987 (million grt)

Source: See text.

Measured in grt, tankers, combination and bulk carriers and general cargo vessels comprised eighty percent of the fleet in 1987. When the figures are converted to cgrt, the share is less than sixty percent for domestically-registered vessels and slightly above that mark for all Norwegian-owned ships. Whereas ferries and passenger vessels comprised six percent of the Norwegian-registered fleet in 1987 by grt, they in fact made up almost a quarter of the fleet by cgrt. Conversely, the largest reduction occurred among tankers and combination carriers; converting cgrt reduces their share from seventy to fifty percent in 1980 and from sixty-two to forty percent in 1985.

The increased sophistication of the Norwegian fleet can be shown by examining the average conversion factor over time (see table 7.11). Perhaps surprisingly, this fell between 1970 and 1975 because of the large influx of mammoth tankers and combination carriers. Moreover, the fleet in 1970 included more than 130 small, general cargo vessels, with an average age of more than fourteen years; by 1975, most had been sold or scrapped. Since the coefficient does not take age into account, the fact that this tonnage was relatively old is not reflected. This implies that the use of cgrt as an indicator of value and replacement cost may actually overestimate the situation in 1970.

The difference between the conversion factors of the Norwegian-registered and Norwegian-owned fleets in 1987 can be explained partly by the fact that large tankers were attractive vessels for flagging-out. On the other hand, some of the other vessels, operating in certain niche trades, may have been better off operating under the relatively prestigious Norwegian flag. This can explain why the owners of passenger vessels, largely operating in the cruise industry, chose to maintain Norwegian registry.

Table 7.11 also shows that the sophistication of the Norwegian fleet, measured using the conversion factors, increased over the period. Measured in grt, the Norwegian-registered fleet declined by more than seventy percent from 1975 to 1987. But if we compare the tonnage in 1975 with the Norwegian-owned fleet in 1987, and use cgrt rather than grt, the reduction in the Norwegian fleet is 27.7 rather than 71.8 percent.

Table 7.11
Tonnage and Conversion Factors, Selected Years, 1970-1987

	1970	1975	1980	1985	1987	1987 int.
Sum group cgrt	10,042,041	12,026,157	10,624,825	8,854,319	4,509,886	8,699,692
Sum group grt	17,476,185	23,656,915	20,786,556	15,227,378	6,662,724	13,098,975
Conversion factor	0.575	0.508	0.511	0.581	0.677	0.664

Source: See text.

This analysis, of course, ignores the age of the vessels. By 1987 the average age was approximately eight and one-half years, about the same as in 1970. The average age peaked at approximately nine years in 1973, but fell to slightly more than seven years in 1981. Thereafter, however, the mean age increased due to the limited new building activity.

The differences in the development of the fleet based on grt and cgrt reflect the structural changes in Norwegian shipping. The state of the industry in 1987 was not as bad as comparisons based upon grt or dwt might indicate. When the fleet peaked in 1977, the majority consisted of large tankers, bulk carriers and combination carriers. Ten years later such vessels were still important, although many of them had been sold or scrapped. But the Norwegian-owned fleet by this time also consisted of a relatively large number of technologically-advanced vessels, operating in niches where the potential for profit was better than in the bulk sector. The value of these vessels per ton was considerably higher than for those that had previously dominated.

The structural transformation implies that the Norwegian fleet became increasingly sophisticated. Parallel with this was increased diversification. Whereas the Norwegian fleet in 1970 largely consisted of general cargo vessels and ships transporting dry and liquid bulk cargoes, the situation in 1987 was characterised by a higher degree of activity in several specialised segments. Norwegian companies owned considerable fleets of passenger vessels, chemical and gas tankers and specialised vessels. Moreover, they had diversified into offshore activities, owning a goodly number of the world's supply vessels and drilling rigs, a fact that has not been accounted for in the previous analysis.

The increased diversification meant that owners were less affected by dramatic changes in a single market segment. The collapse of the tanker market after 1973 had disastrous consequences for Norwegian shipping due to the focus on this segment. The advantage of increased diversification was that although one market might collapse, owners operating other types of shipping would remain relatively unaffected. Norwegian shipping as a whole would thus be less vulnerable. Moreover, several owners operated in niches where price was not necessarily the dominant determinant of competitive ability and where freight rates were more stable.

The transition of the Norwegian fleet was not the result of "creative destruction" and market forces alone – government involvement affected it as well. Governments influenced the trajectory of the crisis in a number of ways, such as through the design of the tax system and support given to shipbuilding. In Norway the authorities' attitude shifted "from benign neglect to active intervention" shortly after the freight market collapse.[26] Throughout the crisis government policies had important implications for the adaptation and transformation of the industry.

Due to the importance of shipping to the economy, the Norwegian government reacted fairly quickly when the market collapsed. The value of ships declined sharply, and Norwegian owners had invested in the types of vessels for which the decline was greatest. Moreover, the fleet was highly mortgaged. The "minimum value clause," which enabled a creditor to demand extra collateral if the value of a vessel fell below a certain level, put a number of shipping companies in a difficult position *vis-à-vis* the financing institutions. To make matters worse, several shipowners encountered liquidity problems as revenues were reduced or non-existent. The combination of low revenues, the minimum value clause and the problematic financial position of Norwegian shipping companies paved the way for government intervention in a sector that had previously been left largely to its own devices. In the summer of 1975 the Department of Trade and Shipping established a committee to advise on how government intervention could ease the immediate financial difficulties of Norwegian shipping companies. The aim was to ensure that owners would not be forced to sell tonnage abroad at the prevailing low prices.

The commission delivered its recommendations within three days of its official appointment. But this is slightly misleading, for in reality it had been working informally for fourteen days and had already discussed two drafts embodying possible solutions. One had been suggested by Haakon Nygaard, who held a crucial position within Norwegian ship financing; he suggested the establishment of a state-owned guarantee institute as well as a na-

---

[26]Helge W. Nordvik, "From Benign Neglect to Active Intervention: Norwegian Government Shipping Policies from the 1970s Shipping Crisis to the Present" (unpublished paper presented to the IAME conference, City University, London, 1997).

tional shipowning company, to be owned by the existing shipping companies and financial institutions to take over Norwegian-owned or -built tonnage.[27] An alternative, presented by the committee's chairman, Hermod Skånland, also suggested a guarantee institute but without the shipowning company. The final report, which received the support of the Ministry of Trade and Shipping, largely reflected his views. The purpose of the guarantee institute was to ensure that Norwegian owners could survive a period of difficult liquidity. Government guarantees were designed to convince financial institutions that the risk of lending to Norwegian owners was lower than it actually was.

On 24 November 1975 the Norwegian Parliament decided to establish Norsk Garantiinstitutt for skip og borefartøyer AS [Norwegian Guarantee Institute for Ships and Drilling Vessels Ltd., GI]. The GI was organised as a limited-liability company in which the government held 6000 of the 10,000 shares. The other shareholders were members of the Norwegian financial community with 2000 shares, and two holding companies owned by Norwegian shipowners and drilling vessel owners, with 1000 shares each. The GI could grant guarantees that would cover interest payments on existing loans; new loans for previously-signed newbuilding contracts; new loans to finance the transfer of vessels or contracts between Norwegian shipowners; and loans covering cancellation fees, lay-up costs and the like.[28] In order to participate, shipowners had to cover ten to twenty percent of the amount. The GI initially was limited to NOK two million, but this was doubled in 1976 once it was realised that the need had been underestimated.[29]

There were two reasons for the inclusion of drilling rigs in the arrangement. The first was that a large number of investors in rigs were shipping companies, which meant that there was a considerable degree of overlap be-

---

[27]Archives of the Norwegian Shipowners' Association, folder 6 B K 75 – Krisen 1975, I, 010175-300675, Haakon Nygaard to Skipsfinansieringsutvalget [Committee Investigating Ship Financing], 17 June 1975.

[28]Norway, Parliament, Stortingsmelding No. 8 (1979-1980), "Om virksomheten i Norsk Garantiinstitutt for skip og borefartøyer A/S i 1978" ["On the Activities of the Norwegian Guarantee Institute for Ships and Drilling Vessels Ltd. in 1978"], 12.

[29]Norway, Parliament, Stortingsproposisjon No. 186 (1975-1976), "Om utviding av statens garantiansvar overfor Norsk garantiinstitutt for skip og borefartøyer A/S" ["Royal Proposition on the Expansion of the Authorities' Guarantee Responsibility towards the Norwegian Guarantee Institute for Ships and Drilling Vessels Ltd."]. A questionnaire sent to potential participants before the arrangement was established indicated a need in the region of NOK 5.4 billion; see Archives of the Norwegian Ministry of Foreign Affairs No. 164.58, Norsk Garantiinstitutt for skip og borefartøyer, box 1, folder 2, 1/10-75-30/11-75, letters from the Norwegian Shipowners' Association [Norges Rederforbund] and Norwegian Association of Drilling Contractors [Norsk Borerigeierforening], 4 November 1975.

tween the two groups. The second was that the rig market had many of the same negative characteristics as the shipping market – over-ordering, an increasing supply surplus and insufficient and insecure demand.

The Norwegian shipping community reacted positively to the establishment of the GI, and 101 companies – three-quarters of them shipowners – entered during the first six months. By the end of 1976 the GI had granted twenty-six guarantees to twenty-three companies.[30] The purpose of the majority granted to the shipping sector was to ensure liquidity, whereas eighty percent of the value of the guarantees granted to rig owners related to long-term financing. The GI granted guarantees for twenty-two large tankers, and at the turn of the decade one-quarter of the Norwegian tanker fleet was owned by GI participants. The participation of shipowners outside the tanker segment was limited, as approximately eighty percent of the guarantees to shipowners went to tanker owners.

In 1981 the guarantees for rig owners were terminated because of a recovery in the rig market without loss for the government and with substantial profit for the companies. The shipping sector guarantees were less successful, largely because the recovery did not materialise. In 1981 the GI suggested that the losses from its engagement in large tankers should be realised, and it suggested a controlled winding-up of the tanker companies by disposing of their tonnage within a year or two. In the end, more than four-fifths of the GI's losses were related to tankers – in particular to a handful of owners with large new building orders and a focus on the spot market.

The GI – and in particular its relationship to Hilmar Reksten and the Aker Group – was (and is) controversial.[31] But the important question is not about hidden motives or foul play but rather about how the arrangement affected the structure of Norwegian shipping. The establishment of the GI may have delayed a necessary and beneficial structural transformation of Norwegian shipping by ensuring that crisis-ridden shipowners could go through the second half of the 1970s without having to sell their ships.

Helen Thanopoulou has claimed that the shipping crisis "coincided with major changes in the distribution of world tonnage among the various

---

[30]Three requests had been rejected, and four of the original participants – two shipowning companies and two rig owners – had withdrawn from the arrangement. Of the twenty-six guarantees granted, fourteen went to the shipping sector and twelve to rig owners.

[31]For an introduction, see Stig Tenold, "The Harder They Come...Hilmar Reksten from Boom to Bankruptcy," *The Northern Marine/Le Marin du nord*, XI, No. 3 (July/juillet 2001), 41-53; and Tenold, "Saving a Sector– But Which One? The Norwegian Guarantee Institute for Ships and Drilling Vessels Ltd.," *International Journal of Maritime History*, XIII, No. 1 (2001), 39-62.

fleets."[32] Due to the fact that government intervention affected the composition of the Norwegian fleet, reducing the amount of tanker and bulk tonnage sold to foreign shipowners, this transition was relatively slow in Norway. After the guarantees for the tankers were terminated, however, the process accelerated.[33]

The creation of a temporary new institution – though more long-lived than originally planned – was the main short-term response to the plight of Norwegian shipping companies. The GI has commonly been viewed as the most important symbol of the government's adaptation to the crisis. Yet the crisis led to more permanent shifts in shipping policy in other areas as well. In the longer term, more drastic measures were necessary, and another response was a shift in governmental policies regarding the use of foreign flags. The depressed state of the market and the shipping companies' economic difficulties were used to justify this change. Consequently, the internationalisation of Norwegian shipping, visible through increased foreign registry, would not have occurred to the same extent if revenues had permitted continued profitable operation under the domestic flag. In this respect, it is evident that the policy changes were a direct consequence of the shipping crisis.

Traditionally, Norwegian legislation had affected the structure of the shipping industry through two mechanisms. By laying down strict rules about ownership, legislation constrained foreign access to the Norwegian register. Moreover, foreign exchange legislation restricted the ability of Norwegians to own ships outside the Norwegian registry.

In the postwar period Norwegian foreign investments were restricted by the so-called "Valutaloven" [Currency Control Act]. As a result of the foreign exchange restrictions, Norwegians were generally not allowed to invest in foreign shipping companies. Shipowners wishing to invest abroad had to apply to the Ministry of Trade and Shipping for exemptions. The authorities claim that the enforcement of this regulation was relatively strict for shipping companies until the late 1970s, but despite such restrictions, some Norwegian-owned vessels were flying foreign flags, either as a result of exemptions or through investments in companies established before the strict currency-control regime was introduced during World War II.

In the 1970s, foreign investments by Norwegian shipowners were low compared with those of most of their competitors. Licenses were not granted if the main reason was to take advantage of conditions such as lower taxes,

---

[32]Helen A. Thanopoulou, "The Growth of Fleets Registered in the Newly-emerging Maritime Countries and Maritime Crises," *Maritime Policy and Management*, XXII, No. 1 (1995), 51.

[33]There were additional reasons for this, such as the increased potential for Norwegian owners to register their vessels in flag-of-convenience (FOC) countries. The GI delayed the flight to FOCs because the participants' vessels were required to fly the Norwegian flag.

cheaper crews, lower social expenses, subsidies or national preferences. The authorities claimed that they had a "relatively restrictive attitude" and that "the traditional practice of allowing registration under foreign flags can be considered restrictive."[34] But a former bureaucrat in the Ministry of Trade and Shipping has claimed that the actual practice had in fact been liberal and was characterised by a great deal of arbitrariness.[35]

In 1975, the government emphasised that "Norwegian shipping, as a main rule, shall be based on vessels flying the Norwegian flag and employing Norwegian crews."[36] It made provisions for the flagging-out of vessels due to high domestic costs, but the scope of this was limited by parliament. It is thus evident that there was some movement towards liberalisation but that there was no majority support for such a change. The shipping crisis, however, made the problems more acute and therefore paved the way for policy changes. In 1979, a parliamentary committee supported temporary "flagging-out" as an alternative to the sale of Norwegian vessels to foreigners.

Table 7.12
Norwegian-owned Tonnage Registered Abroad, Mid-1978 and Mid-1984

	1978	1984
Number of vessels	86	382
Dwt – Norwegian share	1.8 million	8 million
Proportion of Norwegian-registered Fleet	3.8 per cent	25.7 percent

Source: Norway, Stortingsproposisjon No. 46 (1978-1979), "Om støttetiltak for skipsfartsnæringen" ["On Support Measures for the Shipping Industry"], 18; and Stortingsmelding No. 53 (1984-1985), "Om skipsfartsnæringen" ["On the Shipping Industry"], 36.

The large outflow of Norwegian tonnage after the restrictions on foreign registry were eased might indicate a need that previous legislation had been able to staunch. Yet two trends were working in tandem. First, vessels were sold to foreign owners able to operate them more effectively; and second, Norwegian shipowners, encouraged by lower restrictions on manning and wages, transferred parts of their fleets to foreign registries to increase their

---

[34]Norway, Parliament, Stortingsmelding No. 23 (1975-1976) "Om sjøfolkenes forhold og skipsfartens plass i samfunnet" ["On the Conditions for Seamen and Shipping's Role in the Society"], 24; and Norges Offentlige Utredninger (1980:45), "Uteregistrering av skip og skipsfartens egenkapital" ["Foreign Registry of Ships and the Equity of Shipping"], 27.

[35]Gisle Stray Breistein, *Valutaregulering og skipsfart* [*Currency Control and Shipping*] (Bergen, 1984), 83.

[36]Norway, Stortingsmelding No. 23 (1975-1976), 25.

competitiveness. The latter element became particularly important about 1980, when the policy was liberalised. From the late 1970s to the mid-1980s the amount of foreign-registered Norwegian tonnage quadrupled (see table 7.12).

The shift in Norwegian policy was both a response to the crisis and a result of more long-term changes in the maritime sector. An investigation of the parliamentary proceedings illustrates this and indicates when and to what extent the various elements were important.

The liberalisation of Norwegian flag policy in the late 1970s can be considered a direct response to the crisis. The proposition put to parliament referred to previous considerations of the flag issue and emphasised that "the possibilities for profitable operation under the Norwegian flag have been considerably reduced after the parliament's discussion of the Shipping report [1975-1976]."[37] In the recommendation from the extended Committee on Finance and Economic Affairs, "the current state of the shipping industry," i.e., the shipping crisis, was used to justify liberalisation concerning foreign registry.[38] The first steps towards a more liberal shipping policy can thus be directly attributed to the effects of the shipping crisis.

The second stage of liberalisation, however, was not a result of the shipping crisis *per se* but rather a reflection of an increasing disjunction between Norwegian shipping policies and those of other nations. The Ministry of Trade and Shipping emphasised that "[i]n the evaluation of the flag policy, allowances must be made for the fact that an increasing share of the world fleet is registered in countries with free registers." As a result of the policy changes, Norwegian shipowners were allowed to transfer "technically obsolete tonnage" as a way of diversifying their operations.[39]

By the mid-1980s the third stage of liberalisation came into effect when the authorities realised that a "restrictive flag policy will only delay the development towards operation under foreign flags."[40] Accordingly, the government suggested that the procedures should be simplified. The general prohibition was terminated and replaced with laws that granted general access to foreign registries. Instead of disallowing foreign registry, which in certain cir-

---

[37]Norway, Stortingsproposisjon No. 46 (1978-1979), 10.

[38]Norway, Parliament, Innst. S. No. 167 (1978-1979), "Innstilling fra den forsterkede finanskomite om støttetiltak for skipsfartsnæringen" ["Recommendation from the Extended Committee on Finance and Economic Affairs on Support Measures for the Shipping Industry"], 9.

[39]Norway, Parliament, Stortingsmelding No. 52 (1980-1981), "Om skipsfartsnæringen" ["On the Shipping Industry"], 30-31.

[40]Norway, Parliament, Stortingsmelding No. 53 (1984-1985), "Om skipsfartsnæringen" ["On the Shipping Industry"], 41.

cumstances could be waived, the policy became one of tolerance, but with two specific prohibitions.[41] Applications were no longer dealt with on an individual basis but were automatically endorsed unless there were specific reasons for refusal.

The reduction of the Norwegian fleet was dramatic in the period leading up to the changes in flag policy. But three features of Norwegian shipping policy may have delayed the outflow relative to the situation in other OECD countries. First, it is reasonable to assume that some of the vessels that were transferred to foreign registries would have been transferred at an earlier point in the absence of a restrictive flag policy. Second, the Norwegian authorities introduced beneficial tax measures, particularly in connection with the taxation of limited partnerships [*kommandittselskap*], which were available only for vessels flying the Norwegian flag. Third, the existence of the GI contributed to the preservation of large tankers and bulk carriers on Norwegian registry.

---

[41]The exceptions were ships owned by foreign general shipping companies and those owned by limited partnerships in which the investors had no relation to shipping (i.e., tax-motivated investments).

Chapter 8
The Fates of Four Norwegian Shipowners

An analysis of Norwegian shipping at the aggregate level illustrates the basis for, and the implications of, the shipping crisis. But to understand fully the challenges that confronted shipowners it is necessary to look at individual companies and their behaviour both before and after the market crash. The far-reaching, short-term effects of strategic decisions are illustrated through the fates of Knut Knutsen OAS, Hilmar Rekstens Rederi, Rederiet Peder Smedvig and Sig. Bergesen d.y. The analysis is based on the previous business strategy parameters: fleet structure, chartering and contracting.[1]

The actions of the companies in the period leading up to the crisis are paramount in understanding their subsequent fate. The strategy during the hey-day of high levels of contracting and soaring freight rates determined their ability to adapt to market changes, either facilitating, complicating or preventing a sustainable response to the new conditions. The four owners discussed here all had invested in the tanker sector. Although they represent a variety of strategies with regard to fleet structure and chartering, all had huge amounts of tanker tonnage on order prior to the freight market collapse.

Knut Knutsen OAS

The shipping company owned and managed by the Knutsen family had a long history in the liner sector. In the early 1970s it operated two liner services. The older of these, the Knutsen Line, Orient Service, a circular route in the Pacific Ocean calling in America, Asia and Australia, was established in the mid-1950s. To supplement this route, the company launched the Knutsen Line, West Australia-West Pacific, in 1972. Yet it was due to its participation in the tanker sector that Knutsen was particularly affected by the shipping crisis.

Before World War II Knut Knutsen OAS was the third largest shipping company in Norway. The founder, Knut Knutsen, was a distinguished

---

[1]Data on the various fleets have been taken from the Veritas register. This poses problems in connection with ships controlled through nominee companies or those flying foreign flags, such as the flags of convenience (FOCs). These vessels will not be included in the analysis due to the large degree of uncertainty regarding their ownership.

man held in high regard in his hometown of Haugesund.[2] After his death in 1946 the company was divided: one part was merged into Chr. Haaland og Søn, while Knutsen's son, Ole Andreas, continued to manage the remainder under the name Knut Knutsen OAS.

The company dates back to 1896. The first vessel was used to export Norwegian herring. Knut Knutsen OAS is therefore a good example of the manner in which Norwegian shipping evolved from carrying domestic goods to becoming an important participant in the cross-trades. While in Norwegian terms Knutsen's shift occurred relatively late, the timing corresponded well with that of other merchants in Haugesund. The company later diversified, concentrating on tramp shipping in the years before World War I. From the mid-1920s Knutsen invested in whaling ships and a whaling factory. In 1925 Knut Knutsen OAS purchased two 13,000-deadweight ton (dwt) tankers and continued to expand in the tanker sector up to World War II. On the eve of the war, its fleet consisted of eight large tankers and eighteen other vessels.

After World War II the company had two main foci: liners and tankers. Its activities in liners peaked at the end of the 1960s, when this portion of the fleet comprised sixteen ships.[3] In addition, it owned a tanker fleet of 250,000 dwt.[4] From the beginning of the 1970s, the company's tanker expansion accelerated. In October 1970, Knutsen launched the 216,000-ton TT *Elisabeth Knutsen*, which almost doubled the size of the tanker fleet. While the fact that Knut Knutsen OAS was engaged in both liner and tanker operations may suggest an attempt to reduce risk through diversification, it is a good example of a fleet structure that diverged from the firm's strategy and focus.[5]

The changes in the structure of the fleet during the 1970s, when the emphasis was on expanding its involvement in tankers, are apparent in figure 8.1. In 1970 and 1971 five of its liners were lengthened and rebuilt to handle containers. But this was the only major investment in liner operations in the period. While the company retained its liners, it made large investments in

---

[2]Knut Knutsen was the son of shipowner Ole Andreas Knutsen, and the three letters in the name, OAS, represent this – Ole Andreas Sønn (son of Ole Andreas).

[3]Two of these ships, *Bakke Cooler* and *Bakke Reefer*, were employed in the fruit trade and were sold in the mid-1970s.

[4]In the 1950s and 1960s the tanker fleet grew slowly. In 1959 the company owned ten tankers, comprising 235,000 deadweight tons (dwt). Ten years later, the number had been reduced to six, but due to an increase in average size, they amounted to 252,100 dwt.

[5]The company also had a brief interlude in the offshore sector in the mid-1970s; see Per Tronsmo, *Omstilling og organisasjonskultur: lederskap og overlevelsesevne i tre norske rederier* [*Transformation and Organisational Culture: Management and the Ability to Survive in Three Norwegian Shipowning Companies*] (Oslo, 1987).

tankers. The liner fleet was neither renewed nor modernised, despite the fact that the international liner market underwent significant structural and techno-logical changes. In 1978, the average age of the company's liners was almost twenty years, whereas the average age of its tankers was only five.

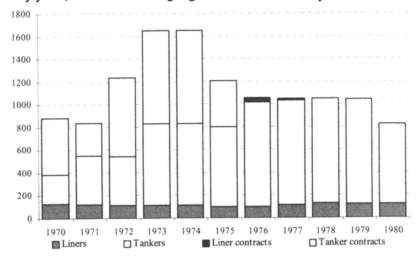

Figure 8.1 Fleet and Contracts, Liners and Tankers, 1970-1980 ('000 dwt)

Source: Knutsen database.

The firm's activities increasingly were directed towards tankers from the beginning of the 1970s. Thor Pande, the company's "heir apparent," was the driving force behind this strategy.[6] The company launched two supertank-ers, TT *Elisabeth Knutsen* and TT *Torill Knutsen*, before the market collapsed. As a result, the tanker fleet in 1973 was almost three times what it had been only a few years earlier. Moreover, Knutsen had ordered two 410,000-dwt tankers, although neither financing nor charters had been secured for them.

On the one hand, the fleet structure of Knut Knutsen OAS indicates a desire to reduce risk through diversification. On the other, the tanker charter-ing suggests a considerable willingness to take risks. The firm usually operated its tankers in the spot market, although medium-term charters were occasion-

---

[6]This strategy was approved by Ole Andreas Knutsen; see Svein Ole Borgen and Martin Spanne, "Investeringsatferd i norske rederier: en casestudie av rederiet Knut Knutsen O.A.S" ["Investment Behaviour in Norwegian Shipowning Companies: A Case Study of Knut Knutsen OAS"] (Unpublished siviløkonom thesis, Norwegian School of Economics and Business Administration, 1982), 17. Pande left the company in 1978 after it had encountered major difficulties due its focus on the tanker sector.

ally taken.[7] One example of such an agreement was the four-year charter signed with Esso for TT *Elisabeth Knutsen*. The revenue from this charter enabled investments in even larger tankers.

The risks undertaken by Knut Knutsen OAS were not confined to the spot-market operation of its own ships. In 1972 it chartered an 85,000-dwt motor tanker from the Japanese shipping company Sanko. This ship, MT *Europride*, was delivered in 1975. When it chartered this vessel, Knutsen had not secured employment for it. Consequently, the charter, like most of the Sanko deals, "proved a considerable financial embarrassment to those owners who were not willing or able to cancel before the tanker market crash."[8]

The Sanko deals are an enticing chapter in the history of merchant shipping in the 1970s. When the Japanese government reorganised the country's shipping industry in the mid-1960s, six companies were given preferential treatment. Sanko was not among them, but it managed to expand by signing large, inexpensive new building contracts, secured through charters with European shipowners. In 1972 more than fifty such contracts were negotiated, and Knut Knutsen OAS was among those ordering vessels through Sanko. The fact that Knutsen entered into such agreements reflects a substantial willingness to "buy" risk and possible profits.

By 1975, all of Knutsen's tankers were laid-up (see table 8.1).[9] According to *Kapital*, the lay-up of TT *Elisabeth Knutsen* and TT *Torill Knutsen* involved annual expenses of NOK thirty or forty million.[10] The sale of a tanker such as TT *Elisabeth Knutsen* in 1979 would have been beneficial, compared with a sale in 1978 or 1980. Knutsen, however, sold it in March as a result of liquidity problems, and the timing was anything but good, since only a few months later prices in the second-hand market were considerably higher.[11] It has been estimated that Knut Knutsen OAS lost NOK 215 million between

---

[7]Tronsmo, *Omstilling og organisasjonskultur*: "The company had gambled enormously on the tanker operations. It operated in the spot market, like it always had."

[8]Peter Stokes, *Ship Finance: Credit Expansion and the Boom-Bust Cycle* (London, 1992), 81.

[9]At this point the tanker fleet consisted of the two supertankers, TT *Elisabeth Knutsen* and TT *Torill Knutsen*, as well as the motor tankers *John Knutsen* and *Anna Knutsen*. The latter two were sold to owners registered in the Cayman Islands in October 1975 for US $11.5 million.

[10]*Kapital*, No. 14 (1975).

[11]TT *Elisabeth Knutsen*, built in 1970, was sold for US $6.45 million to Korean interests. In August, the slightly smaller TT *Jarmona*, built in 1968, was also sold to Koreans for US $9 million.

1975 and 1979. This can be compared with a bottom line of NOK fifty-five million in 1973 and NOK sixty million in 1974.[12] The relatively risky strategy, with a considerable share of the tonnage in the spot market, was the main cause of the extremely large fluctuations in operating profits.

Table 8.1
Knut Knutsen OAS: Employment of Large Tankers, 1970-1980[13]

	TT *Elisabeth Knutsen* (216,000 dwt)	TT *Torill Knutsen* (285,000 dwt)	TT *Hilda Knutsen* (410,000 dwt)	*Kawasaki 1215* (410,000 dwt)
1970	Delivered			
1971	TC Esso			
1972	TC Esso		Ordered: 270,000 dwt	
1973	TC Esso	Delivered	Converted to 410,000 dwt	Ordered
1974	TC Esso	Laid-up (some trips)		
1975	Laid-up	Laid-up		Converted (two liners)
1976	Laid-up	Laid-up	Delivered	
1977	Laid-up	Laid-up	Laid-up (some trips)	
1978	Laid-up	Laid-up until August	Laid-up (some trips)	
1979	Sold	Spot market	Spot market	
1980		Laid-up (some trips)	Laid-up (some trips)	

Sources: Knutsen database; Svein Ole Borgen and Martin Spanne, "Investeringsatferd i norske rederier: en casestudie av rederiet Knut Knutsen O.A.S" ["Investment Behaviour in Norwegian Shipowning Companies: A Case Study of Knut Knutsen OAS"] (Unpublished siviløkonom thesis, Norwegian School of Economics and Business Administration, 1982), appendix 1; and *Kapital*, various issues.

Between 1970 and 1980, all the new building contracts signed by the company were for tankers. Because new building orders are a better measure of the company's strategy than fleet structure, it is evident that the focus had

---

[12]Tronsmo, *Omstilling og organisasjonskultur*.

[13]The table is based upon figures from the Knutsen database, as well as information from Borgen and Spanne, "Investeringsatferd i norske rederier," appendix 1; and *Kapital*, various issues.

shifted from liners to the relatively risky tanker market.[14] The turbine-driven tanker *Elisabeth Knutsen*, delivered in 1970, proved a very profitable investment. The company's second supertanker was also launched before the oil price increase. Due to the favourable conditions, the increased investments in tankers led to high profits in the early 1970s; this encouraged the company to order more large tankers.

In late 1972, Knut Knutsen OAS ordered a 270,000-dwt tanker at Kawasaki. In the summer of 1973 this contract was converted to 410,000 dwt, and early in the autumn the company ordered an identical ship.[15] The contract was signed without secure financing or employment.[16] Shortly after its signing, oil prices increased, and the tanker market collapsed.

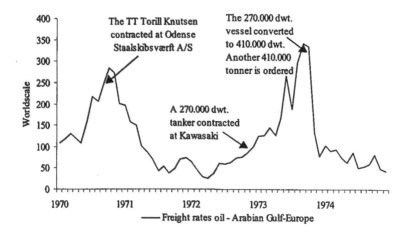

Figure 8.2:   Freight Rates and Contracting by Knut Knutsen OAS, 1970-1974

Source: See text.

[14]According to the lists in the *Norwegian Shipping News*, Knut Knutsen OAS contracted a 14,000-dwt vessel at VEB Mathias-Thesen Werft. This is a confusion between Knut Knutsen OAS and the Oslo-based ship owner Knut A. Knutsen.

[15]This contract was registered in the *Norwegian Shipping News* for the first time in January 1974 and is not included in the survey of new building contracts from September 1973.

[16]This contract would haunt Knutsen until 1982, when the company had to come to terms with an agreement with the Guarantee Institute (GI) and the Japanese yard; see *Økonomisk Rapport*, No. 16 (1987), 86.

Knut Knutsen OAS signed new building contracts when its liquidity was good or when freight rates were high. Since the company operated most of its tankers in the spot market, these conditions usually occurred concurrently. It has been claimed that "analysis was banned in periods were the tanker market was good."[17] In the same manner, vessel sales were sometimes motivated by liquidity problems. Accordingly, decisions to purchase and sell ships were not always part of a conscious strategy but rather were influenced by external events, such as market conditions and the company's liquidity.

By the time that the new buildings ordered in 1972 and 1973 were delivered, the bottom had fallen out of the market. TT *Hilda Knutsen*, building number 1215 at Kawasaki, was originally scheduled for launch in late 1974 or early 1975. The ship was delivered in April 1976, and had seven charters before September 1978. Building number 1228, *Hilda Knutsen*'s sister-ship, was never delivered. In 1975 the company entered into negotiations with the yard to convert the contract to two liners. The idea behind the conversion was to spare Knutsen from having to pay the cancellation fee or being forced to operate the ship in a depressed tanker market. Due to the size of the conversion fee and the prevailing exchange rates, the two liners became twice as expensive as they would have been if ordered in a conventional manner.[18]

Primarily because of the orders for two large tankers without employment, the company's losses amounted to more than NOK 200 million between 1975 and 1979. The result was substantial economic difficulties. Yet the ordering of ever-larger tanker tonnage was an important part of the company's strategy. The sharp expansion began with the ordering of TT *Elisabeth Knutsen* in the late 1960s and ended abruptly when the tanker market collapsed.

The British Hambros Bank was originally Knutsen's most important financial partner. After the freight market collapse, Hambros Bank suffered large losses on the financing of two major Norwegian shipowners, Hilmar Reksten and Hagbart Waage. As a result, Hambros would not make Knut Knutsen OAS an acceptable offer when the final part of the financing for *Hilda Knutsen* was to be arranged in 1975. The company therefore had to seek alternatives, and the Guarantee Institute (GI) became the guarantor for the loans, which were taken out not only for *Hilda Knudsen* but also to provide liquidity. A guarantee for a liquidity loan in late 1981 was motivated by the fact that:

> the discontinuation of the Guarantee Institute's involvement in the tankers would lead to the winding-up of the company's additional activities...As the company was one of the few remaining small Norwegian liner companies, the winding-up

---

[17]Unnamed employee quoted in Tronsmo, *Omstilling og organisasjonskultur*.

[18]Borgen and Spanne, "Investeringsatferd i norske rederier," 47.

would be a further step in the direction of reduction of Norwegian liner activities.[19]

The neglected liner sector inadvertently came to the rescue. While the guarantee subsequently was rejected by the Ministry, the GI and Knutsen's main creditors became strongly involved in the company's operations.[20]

Knut Knutsen OAS was originally involved in both tankers and liners. Motivated by expectations of large profits, Knutsen expanded in the tanker sector, at the same time neglecting the need for modernisation and renewal of the liner fleet. The result was that when the tanker market collapsed, the company found itself with the bulk of its resources in a market that suffered from overcapacity and depressed rates. Ironically, the tanker expansion meant that Knutsen became less competitive in the relatively secure liner sector due to lack of investment.[21] The story of Knut Knutsen OAS illustrates how a focus on the tanker market could have disastrous consequences. The expansion enabled the company to reap large profits in the first part of the 1970s. But the ordering of two Ultra Large Crude Carriers (ULCCs) without financing or secure employment proved extremely unfortunate when the market collapsed. As a result of the contracting in 1973, the company acquired a supertanker that alternately was laid-up or earned abysmal rates, as well as two liners that had become very expensive. Moreover, it was forced to sell parts of the fleet due to acute liquidity problems. The fate of the firm also illustrates the changes wrought by the shipping crisis. By the mid-1980s, after almost ninety years at the helm, the Knutsen family was forced to ship the oars.[22] Knutsen OAS today operates a fleet of some thirty-five vessels, mainly shuttle tankers, product carriers and chemical tankers. The problems of the 1970s have left their mark, and today the company generally operates on long-term charters.

---

[19]Norway, Parliament, Stortingsproposisjon No. 21 (1982-1983), "Om bevilgning til utbetaling under garantier stillet av Norsk garantiinstitutt for skip og borefarttøyer AS for lån til rederiet Knut Knutsen OAS" ["On the Appropriation of Payment with Regard to Guarantees Granted by the Norwegian Guarantee Institute for Ships and Drilling Vessels Ltd. for Loans to the Shipowning Company Knut Knutsen OAS], 1-2.

[20]The GI, together with Manufacturers Hanover Trust and Bergen Bank, forced Knutsen at the end of the 1970s to recruit a managing director with no previous relation to the company.

[21]Because the increasing preferential treatment of domestic firms altered the conditions in the liner market from the mid-1970s, it is possible that larger liner investments would only have delayed the economic difficulties.

[22]The basis for the withdrawal can be found in Bergen Bank's proposal to avoid bankruptcy referred to in Stortingsproposisjon No. 21 (1982-1983), 2-3.

Hilmar Rekstens Rederi

In the early 1970s Hilmar Reksten was one of the most famous characters in international shipping. This reputation was primarily due to his risky strategy and the large profits earned on a series of speculative deals. "The mysterious Norwegian Hilmar Reksten" was called a "gambler" and an "archetypal entrepreneur," while others referred to him as "risk-loving" and "the shipowner who was willing to break law and justice to accomplish his goal [of] becoming the largest tanker owner in the world."[23] There are some difficulties in assessing Reksten's fleet because he had considerable assets abroad, out of the reach of Norwegian authorities. Among other things, it has been alleged that Reksten, together with P&O, controlled Associated Bulk Carriers, one of the world's largest dry bulk companies.[24] Due to the large degree of uncertainty about Reksten's foreign assets, only the ships listed as owned by Reksten in *Veritas* are regarded as part of his fleet in this analysis.

Even if an assessment of Reksten's fleet is fraught with uncertainty, his strategy is easy to discern. Although known as a relatively shy person who seldom gave interviews, he explained and defended his strategy and business philosophy in public on several occasions. These sources are important to understand certain elements of Reksten's market behaviour and contracting.[25]

Hilmar Reksten embarked on a career in shipowning in the interwar period after studying in Germany. Before he went to Germany, he speculated in *marks* using a loan of NOK 5000 from a Bergen shipowner that was intended to finance his studies. Unfortunately for Reksten, the devaluation of the German currency rendered his investment nearly worthless in only four months. Reksten sometimes used this story to illustrate the uncertainty of the currency market, and the incident might explain his fear of exchange rate

---

[23]Quotes from *The Sunday Times* (London), 12 March 1978; Harald Espeli, *Industripolitikk på avveie – Motkonjunkturpolitikken og Norges Industriforbunds rolle 1975-80* [*Industrial Policy Gone Astray – The Counter-cyclical Policy and the Role of Norges Industriforbund 1975-1980*] (Oslo, 1992), 55; and Erling Borgen, *Hilmar Rekstens eventyr* [*The Adventures of Hilmar Reksten*] (Oslo, 1981), 7.

[24]See Erling Dekke Næss, *Autobiography of a Shipping Man* (London, 1977), 233; and *Norwegian Shipping News*, No. 12 (1973), 402 and No. 15 (1973), 487. Reksten claimed that he was only involved in the management of the vessels on behalf of the real owners who preferred anonymity.

[25]The advantage of such sources is that they make it possible to examine the evaluations. The problem with using them in connection with Reksten is that even after being proved wrong he claimed that he acted rationally. He also tended to blame circumstances over which he had no control for his failures.

losses in connection with long-term charters.[26] Reksten began by operating his own and chartered vessels on the North American Great Lakes. In 1938 he ordered his first tanker at the Swedish yard Kockums Verft.[27] After World War II, Reksten questioned the government's licensing policy, which he saw as discriminatory against owners wishing to invest in large tankers.[28]

Despite feeling obstructed by the government, Reksten continued to expand into tankers. In the early 1960s he ordered seven turbine-driven tankers for delivery between 1965 and 1967.[29] The order for TT *Julian* at Aker/Stord marked the beginning of a close association between Reksten and the Aker Group. In the years 1964-1972, Reksten contracted with the yard for twelve Very Large Crude Carriers (VLCCs), worth more than NOK two billion.[30]

In addition to his own shipping firm, Hilmar Reksten had economic interests in a variety of Norwegian companies, both in shipping and elsewhere. He also owned several properties and intended to build a shipyard on one of them.[31] He also owned a plethora of stocks and properties abroad. When Reksten encountered liquidity problems in the mid-1970s, the authorities decided to buy his Norwegian shares, which were valued at NOK 200 million.[32]

Reksten's emergence as one of the world's leading tanker owners was the result of a philosophy that rested on three principles: a focus on tankers; the use of the largest and most efficient vessels; and a reliance upon operations

---

[26]*Norwegian Shipping News*, No. 7B (1971), 49.

[27]In *ibid.*, No. 1 (1976), 20, Reksten claimed that he originally wanted to contract this ship in Norway but was surprised to learn that domestic yards were unable to build the kind of vessel he wanted.

[28]*Ibid.*, No. 7B (1971), 53; and Hilmar Reksten, "Noen ideer om konkurransevilje og risikomomentet under strukturendringene i norsk tankskipsfart" ["Some Ideas about the Willingness to Compete and the Risk Aspect during the Structural Changes in Norwegian Tanker Shipping"] (Kristofer Lehmkuhl Lecture, Norwegian School of Economics and Business Administration, 1971), 7-16.

[29]The vessels were contracted in Germany, Great Britain, Sweden and, in the case of TT *Julian*, Norway.

[30]Rederiaksjeselskapet Hadrian, *Driftsberetning og Resultat* [*Annual Report*] (Oslo, 1972), 7.

[31]*Ibid.* (Oslo, 1977), 12. The yard was to build tankers in the region of 800,000 dwt; see *Norwegian Shipping News*, No. 15 (1973), 485.

[32]Norway, Parliament, Stortingsproposisjon No. 187 (1974-1975), "Om fullmakt til kjøp av aksjer i norske selskaper" ["On the Authorisation of the Purchase of Shares in Norwegian Companies"].

in the spot market rather than long-term contracts. Early in the 1970s Reksten claimed that he controlled the largest privately-owned independent tanker fleet in the world.[33] This consisted mainly of turbine-driven supertankers, which in 1970 amounted to more than one million dwt; he also had a similar sized "fleet" in the form of new building contracts at German and Norwegian yards. Reksten had invested heavily in tankers, and after 1970 his fleet consisted almost exclusively of these vessels. In 1975, Reksten received the gas-tanker GTT *Lucian* from Moss-Rosenberg Verft, allegedly after signing the US \$40 million contract for the ship on a napkin.[34] Despite the inclusion of this vessel, more than ninety-five percent of his fleet consisted of oil tankers.

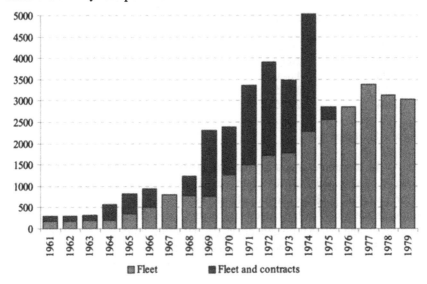

Figure 8.3: Reksten's Fleet and Contracts, 1961-1979 ('000 dwt)

Note: Vessels ordered in Reksten's name but intended for the company of his son, Audun Reksten; those owned by Reksten through foreign companies; and ships chartered from other shipping companies are excluded.

Source: Reksten database.

One conspicuous feature of Reksten's operations was the large number of contracts relative to the size of the fleet (see figure 8.3). Although Norwegian shipowners had entered into a disproportionate share of the world's

[33]Hilmar Reksten, *Opplevelser* [*Experiences*] (Oslo, 1979), 29. The assertion is correct if only vessels without charters are deemed "independent."

[34]Borgen, *Hilmar Rekstens eventyr*, 88.

new building contracts, Reksten's orders, relative to his existing fleet, were even larger than those of his Norwegian colleagues. One important reason was that Reksten ordered ships of ever-increasing size (see table 8.2). Fleet structure is crucial in explaining why Hilmar Reksten was hit extremely hard by the shipping crisis. First, his fleet consisted almost exclusively of tankers. Second, the vessels were large, and it was the freight rates and values of the largest tankers which deteriorated the most during the crisis.

Table 8.2
Average Size of Hilmar Reksten's Tankers and Tanker Contracts, 1970-1974

	Average Size, Existing Fleet (dwt)	Average Size, New Build Contracts (dwt)
1970	90,500	221,000
1971	100,000	265,000
1972	114,000	275,000
1973	196,500	285,000
1974	227,500	352,500

Source: See figure 8.3.

In addition to operating his own large fleet, Reksten increased both his potential profits and risks by chartering four vessels from other shipowners. When he encountered economic difficulties, Reksten did not have the financial resources necessary to honour the agreements, and the charters were annulled. As a result, the original owners demanded US $300 million in compensation.[35]

Hilmar Reksten had a clear chartering strategy: only in particular circumstances would he commit his vessels to charters. This strategy was based upon the fact that he assessed the risk of operating in the spot market differently from his competitors. He claimed that his "chartering policy...was based on commercial reasoning and not on gambling. I could not afford to gamble."[36] In his autobiography, *Opplevelser [Experiences]*, he characterised "the four times I hit the bull's eye in the spot market" as the main achievements in his shipping career. In 1968, MT *Julian* was committed to a five-voyage charter, and the ship's profit during these nine months far exceeded its initial price. Four years later, MT *Octavian* was chartered for five years, quite unusual given his general chartering policy. Reksten claims that this "is an example

---

[35]See *Kapital*, No. 18 (1978), 21. For an introduction to the saga surrounding the chartered vessels, see Stig Tenold, "The Harder They Come...Hilmar Reksten from Boom to Bankruptcy," *The Northern Marine/Le Marin du nord*, XI, No. 3 (July/juillet 2001), 41-53.

[36]Reksten, *Opplevelser*, 169.

that I committed vessels to charters when I felt that it was right."[37] The largest block-closure included twelve of Reksten's fifteen vessels, which were chartered to British Petroleum. This agreement, made in 1970, involved fifty-nine voyages around the Cape.[38] Reksten's spot market victories culminated in the closure of the VLCC TT *Kong Haakon VII* at Worldscale 400. Reksten claimed that this was "a rate the world has never seen, neither before nor after."[39] The net profit from this trip was NOK forty-two million, implying that forty percent of the ship's building price was repaid by a voyage that lasted a mere sixty-eight days.

On several occasions Reksten emphasised that he operated his vessels in the spot market because he regarded this as the sector with the lowest risk. Fear that shifts in exchange rates or costs would reduce his profits made him consider the charter market as a risky alternative:

> the risk associated with this assessment of the market conditions is, in periods of depression, not unreasonably large for experienced shipowners with a practised organisation and competitive tonnage, which in periods of increasing demand also can reap unreasonably large profits.[40]

One week before the Organisation of Petroleum Exporting Countries (OPEC) began its concerted effort to raise oil prices, Hilmar Reksten claimed in a public interview that "[a]gain I am in the fantastic position that all my vessels are by the goal line, unchartered and free."[41] The "fantastic position" was reflected in the closure of TT *Kong Haakon VII* at Worldscale 400. But immediately after the ship finished its trip, it was laid-up for four months. The vessel then undertook one voyage at rates that did not cover costs.

Table 8.3 is a survey of one of Reksten's companies, RA Hadrian, and the employment of some of its tankers. Reksten's spot-market policy proved to have dramatic consequences, and at one point in 1975 all his tankers were laid-up. A few weeks before the tanker market collapsed, Reksten alleg-

---

[37]*Ibid.*, 168.

[38]It has been claimed that Reksten made considerably larger profits from these voyages than he declared to the Norwegian authorities, and that this revenue enabled him to buy his share of the Zapata-Næss fleet.

[39]Reksten, *Opplevelser*, 169.

[40]Reksten, "Noen ideer," 21.

[41]Interview with Hilmar Reksten, 9 October 1973, reprinted in *Norges Industri*, 22 October 1973.

edly rejected an offer of long-term charters for several of his vessels because
of a difference of ten cents per ton between the charterer's offer and his demand.[42] The result was that he ended up with large parts of his fleet in lay-up
and new building contracts for even more unchartered vessels.

Table 8.3
RA Hadrian's Tankers and Employment, 1973-1978

	TT *Vespasian* (285,700 dwt)	TT *Hadrian* (219,720 dwt)	TT *Octavian* (219,720 dwt)	TT *Cyprian* (285,400 dwt)
1973		Spot market, inactive from October	Two voyages, six-voyage charter to August 1974	
1974	Delivered June, three voyages	Spot market until November, laid-up	Laid-up from August	
1975	Laid-up from February	One voyage, laid-up	Laid-up	Delivered
1976	One voyage, Laid-up	Laid-up	Laid-up, four voyages	Laid-up
1977	Two voyages, laid-up	Laid-up	Two voyages, laid-up	Three-year charter
1978	Two voyages, laid-up	Laid-up	Laid-up until November	Three-year charter

Source: Rederiaksjeselskapet Hadrian, *Driftsberetning og Resultat* [*Annual Report*],
                various years.

Hilmar Reksten usually ordered new buildings in blocks, placing orders for a series of vessels at the same time (see figure 8.4). Early in 1970, he
ordered three ships, totalling 850,000 dwt, from the Aker group's Stord
yard.[43] Less than a year later, two 283,000-dwt ships were added. The final
and biggest block was signed shortly before the market broke down when Reksten ordered four 420,000-dwt tankers.[44] The large ordering of VLCCs and

---

[42]Borgen, *Hilmar Rekstens eventyr*, 84.

[43]Reksten, *Opplevelser*, 11-24, is a subjective introduction to the problems
surrounding his contracts with Aker from the beginning of the 1970s. An alternative
version of these events can be found in Onar Onarheim, *Min tørn* [*My Bout*] (Oslo,
1984), 177-182. Figure 8.4 is based on the dates that Reksten presented for his contracts rather than those in the *Norwegian Shipping News*' list.

[44]These contracts were not registered in the *Norwegian Shipping News*' list of
vessels contracted by Norwegian owners, published in September 1973. Onarheim, *Min*

ULCCs, usually at fixed prices, meant that Reksten's prices were favourable compared with those of his competitors.[45]

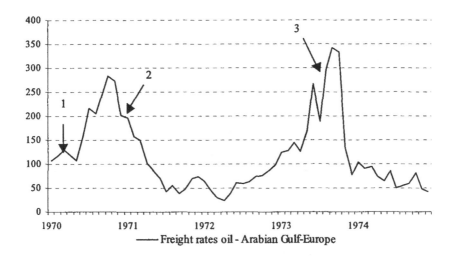

Figure 8.4: Freight Rates and Tanker Contracting, 1970-1974

Source: Reksten database.

The massive contracting of consistently larger tonnage was an important element in Reksten's strategy. The effect of the focus on large vessels was clearly visible when we compare his fleet and new building contracts with the Norwegian and international averages (see figure 8.5).

All of Reksten's contracts were entered into without the security of long-term charters. Hilmar Reksten experienced the dangers inherent in this contracting policy as early as 1972, when TT *Hadrian* went directly from the Aker/Stord yard into lay-up. Two years later, as a result of the spot-market policy, the net profit of the same ship was more than NOK fifty million. In 1974 Reksten's turbine-tanker contracts at Aker/Stord amounted to approximately three million dwt, and these ships were to be launched in a market suffering from overcapacity and bunker prices that favoured motor tankers. At the same time, it was evident that the value of the company's fleet was deteriorating, something that would create difficulties with the creditors. The solution was to cancel some of the contracts with Aker/Stord.

tørn, 56-65, has a discussion of the negotiations from June 1973 which ended in the contracting of the four tankers.

[45]*Kapital*, Nos. 7-8 (1973), 41, claimed that there were large hidden reserves in the Reksten fleet.

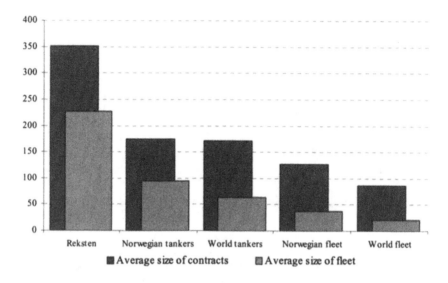

Figure 8.5:   Average Size of Fleets and Vessels on Order, 1974 ('000 dwt)

Source: The figures for Reksten are from the Reksten database. The figures for the Norwegian and international fleets and contracts are from Fearnley and Egers Chartering Co. Ltd., *Review 1975* (Oslo, 1976), which includes tankers and bulk carriers of more than 10,000 dwt and other vessels over 1000 grt.

Reksten had difficulty paying the instalments on four tankers as early as April 1974. He claimed that the lack of payment was a protest against the fact that his previous instalments had been used for purposes other than his own ships. At the same time, he indicated that he intended to cancel several of the contracts because the yard would be unable to deliver the vessels on schedule or to secure financing in the manner stipulated in the contract. In September 1974 Aker decided that the contracts for the four 420,000-tonners should be cancelled and went to court to secure compensation for its losses. Three months later, Reksten cancelled the remaining contracts, citing foul play from the yard as the reason. As a result of the annulment of the contracts, an arbitration court ordered Reksten to pay Aker NOK 320 million.[46] In addition, he had to pay more than NOK 200 million in cancellation fees for the two 283,000-dwt vessels.[47] In the end, due to his eager contracting in 1972 and

---

[46]The NOK eighty-six million already paid was deducted.

[47]A total of NOK 107 million was deducted from the total sum due to instalments previously paid by Reksten.

1973 Reksten had to pay NOK 520 million for vessels he never received. Even for a company with a more sustainable chartering strategy, it is doubtful that expenses of this magnitude would have been manageable. Unfortunately, Reksten's liquidity problems coincided with a sharp reduction in the value of his assets. Estimates from a Norwegian shipbroking company show that the value of a standard 200,000-dwt tanker decreased by eighty percent between 1973 and 1975.[48] Reksten has estimated that the value of his fleet fell by more than NOK two billion, or by two-thirds, from January 1974 to September 1975.[49] His net wealth fell from NOK 2.7 billion to red figures over the same period.

Hilmar Reksten's ordering was unorthodox. When the vessels were ordered, the only guarantee was Reksten's signature; neither charter agreements nor bank guarantees were necessary.[50] The reason for these unconventional deals was Reksten's close relationship to Hambros Bank.[51] New building contracts were financed mainly by yard credits, originating with state-run agencies, but additional financing was provided by Reksten personally or by Hambros. There are several indications that Hambros shared Reksten's opinion that operating tankers in the spot market was the best strategy.[52]

Hambros Bank played a crucial role in financing shipbuilding for Norwegians, and several of the shipowners who contracted large parts of their tonnage from Norwegian yards had Hambros Bank as their main banker.[53] After the tanker market collapse, the bank's losses on loans to Norwegian owners, particularly Hilmar Reksten and Hagbart Waage, were enormous. According to the *Norwegian Shipping* News, the bank had to write off £8.6

---

[48]Fearnley and Egers Chartering Co. Ltd., *Review 1976* (Oslo, 1977), table 19.

[49]Reksten, *Opplevelser*, 175.

[50]*Ibid.*, 7. In an interview with the *Norwegian Shipping News*, No. 7B (1971), 59, Reksten asserted that "if one becomes too involved with finances one begins to be afraid and to be dependent upon the banks. I do not recommend it."

[51]Borgen, *Hilmar Rekstens eventyr*, 107, claims that Hambros Bank "secured their own business by taking security in Reksten's foreign assets, but in Norway let the authorities compensate through guarantees granted by Norsk Garantiinstitutt for Skip og Borefartøyer AS."

[52]See, for example, Rederiaksjeselskapet Hadrian, *Driftsberetning og Resultat* [*Annual Report*] (Oslo, 1975), 5.

[53]See Jan A. Hammer and Arvid Høy, *Norsk skipsfart og dets lånekilder* [*Norwegian Shipping and Its Sources of Finance*] (Bergen, 1983), 36.

million in 1975/1976 due to its involvement with Reksten.[54] According to the bank's annual reports, the losses amounted to £4.6 million in 1976, an undisclosed amount transferred from the bank's inner reserves in 1978 and more than US $60 million in 1982.[55]

In the autumn of 1975, Reksten was teetering on the verge of bankruptcy. The transfer of funds from a mysterious foreign company and an agreement between the Norwegian government, the Aker group and Hambros Bank were necessary to secure the continuation of Reksten's companies. Reksten was also one of the first Norwegian shipowners to require the assistance of the GI, and a large portion of its resources were channelled into his companies. Indeed, it has been claimed that at least part of the motivation for establishing the GI was to save Reksten from bankruptcy and to the close links between Reksten and the Aker group.[56] Nonetheless, the entry of the GI signalled the end for Hilmar Reksten. A new company, RA Trajan II, was established to take over the vessels. Moreover, the family's eighty percent shareholding in RA Hadrian was transferred to the GI without compensation. The vessels were to be managed by Reksten's adopted son, Johan. Although Hilmar Reksten was allowed to remain on the board of RA Hadrian, he refused to sign the annual report and chose not to participate in board meetings because "he questioned the legal basis of the agreements with the Aker group and the Guarantee Institute."[57] In the period 1976-1980 the annual losses were in the region of NOK 270-300 million. The accumulated losses increased from NOK 131 million in 1976 to NOK 1275 million by the end of 1980. In late 1981 the GI disbursed NOK 871 million to cover the guarantees granted to Reksten's companies. On 10 May 1982 RA Hadrian was wound up after almost fifty years of operation: its share capital of NOK 17.4 million was lost, and accumulated losses

---

[54]*Norwegian Shipping News*, Nos. 13/14 (1978), 13.

[55]Hambros Bank, *Report of the Directors and Accounts 1976* (London, 1977), 3; and *Report of the Directors and Accounts 1978* (London, 1979), 5. The bank reached two expensive settlements with the Norwegian government in 1982 and 1995; see Stokes, *Ship Finance*, 41-42; and Tenold, "Harder They Come."

[56]This motive has been presented both by the Guarantee Institute's barristers and by Haakon Nygaard, one of the instigators behind the institution and later its managing director. See *Aftenposten*, 29 October 1992. Nygaard's claim can be found in Norway, Parliament, "Rapport til Stortinget fra den granskningskommisjon i Reksten-saken som ble oppnevnt ved Stortingets vedtak 20. juni 1985" ["Report to the Storting from the Investigative Committee in the Reksten Case Established by Royal Resolution 20 June 1985], 9 February 1988, 199.

[57]Rederiaksjeselskapet Hadrian, *Driftsberetning og Resultat* [*Annual Report*] (Oslo, 1976), 32.

amounted to NOK 567 million. RA Trajan II continued its activities with Johan Reksten and Hambros Bank as the main shareholders.

Since Hilmar Reksten is often regarded as the personification of the shipping crisis, he deserves more attention than other shipowners.[58] Moreover, his demise was as controversial as his business policy. It included, among other things, several lengthy court cases, the biggest civil action in Norwegian legal history and threats of the impeachment of government members involved in the case. In 1979 Reksten was charged with eight counts of tax evasion and exchange-control violations. Although he was acquitted on seven of the charges, he was fined NOK one million for offences in connection with the sale of a tanker in 1973. He condemned the criminal proceedings against him in a letter to Norwegian Prime Minister Odvar Nordli, complaining that "[e]ven my iron constitution could not endure the strain."[59] He died in Bergen in July 1980 at the age of eighty-two. His estate was declared bankrupt by the authorities in November 1980 due to defaults on tax payments. In January 1991, more than ten years after the fall of his empire began, the Norwegian newspaper *Aftenposten* claimed that "[t]he Reksten-case is definitely approaching an end."[60] This was perhaps too hopeful, for legal cases relating to his estate have been before the Norwegian judiciary ever since. Indeed, even the Norwegian national budget for 2006 noted that "the Guarantee Institute has been carried on due to a number of lawsuits related to the Reksten complex."[61]

In his autobiography Reksten proposed a scheme for cooperation among Norwegian tanker owners.[62] Two strategic elements were crucial: that no long-term charters should be signed due to the risks of exchange rates and inflation, and that investments should be in turbine-driven supertankers. This collaboration never materialised, although Reksten certainly followed his own suggestions. Operations in the spot market proved very profitable when demand for tanker transport was increasing, but when the freight market collapsed it became impossible to secure worthwhile employment for ships with-

---

[58]For a more detailed assessment of Hilmar Reksten's strategy and fate, see Tenold, "Harder They Come."

[59]Archives of the Norwegian Shipowners Association, folder 6 B K 75 – Krisen 1975 XII: 011178-300679, Hilmar Reksten to Prime Minister Odvar Nordli, 26 March 1979.

[60]*Aftenposten*, 24 January 1991.

[61]Norway, Parliament, Stortingsproposisjon No. 1 (2005-2006), "Statsbudsjettet – Nærings- og Handelsdepartementet" ["The National Budget – Proposition from the Ministry of Trade and Industry"], 37.

[62]Reksten, *Opplevelser*, 163-167.

out charters. Investments in large turbine tankers were also extremely profitable as long as demand increased faster than supply. But after the oil price increase it was this segment of the market that experienced the most precipitous decline, both in freight rates and the value of vessels.

On 9 October 1973 Hilmar Reksten proclaimed that he was "in a fantastic position" because he had more tonnage available in the booming spot market than any other shipowner. A few years later he set a more dubious record: no shipowner had more laid-up tonnage. As Peter Stokes put it, "[t]he highest flyer in 1973, Hilmar Reksten, was the first to plunge earthwards...Reksten, whose ambitions in 1973 appeared to know no bounds, was a classic over-reacher, who sooner or later was bound to come to grief."[63]

Rederiet Peder Smedvig

Rederiet Peder Smedvig was a relatively diversified firm involved in a variety of businesses. The Smedvig family's companies were concentrated originally in shipping and canned goods. When the founder, Peder Smedvig, died in 1959, his son Torolf took control. Torolf Smedvig has on several occasions been characterised as a "typical" entrepreneur, and he extended the company's interests to include herring oil, property development, tourism and in particular offshore oil-related activities.[64] This short analysis focuses on the firm's activities in shipping, with a particular emphasis on tankers. Rederiet Peder Smedvig differs from the other businesses discussed due to its considerable diversification outside shipping. Nonetheless, the company's fate in the tanker sector illustrates how a few untimely decisions can have large consequences.

When Peder Smedvig decided in 1915 to retire from a career at sea, he already had shares in several vessels.[65] The company's first tanker, MT *Glittre*, was purchased twenty years later. It was originally intended for the

---

[63]Stokes, *Ship Finance*, 35 and 37.

[64]Gunnar Nerheim and Bjørn S. Utne, *Under samme stjerne – Rederiet Peder Smedvig 1915-1990* [*Under the Star – The Shipowning Company Peder Smedvig 1915-1990*] (Stavanger, 1990), 179. This book is an excellent introduction to the history of Smedvig's companies and a fine example of the potential of shipping company histories. My analysis is to a considerable extent based on information from the book.

[65]*Ibid.*, 37, claims that Peder Smedvig "joined the ranks of the shipowners" in 1915. This can be explained by the fact that this was the first time Smedvig was more than a passive investor. From 1915 onwards he acted as managing director and chairman of the board.

spot market but in 1937 was chartered to Standard Oil at a very good rate.[66] During World War II, Smedvig lost four of its five vessels.

Peder Smedvig's son, Torolf, had joined the shipowning company in 1939. After he took the reins in 1959, he launched his diversification scheme. Of crucial importance for its maritime interests, the company was among the pioneers in the Norwegian offshore oil industry, and in 1971 Torolf Smedvig was among the first Norwegian shipowners to contract for oil rigs. Even earlier, however, the company had entered the offshore industry through investments in supply ships and onshore supply bases.

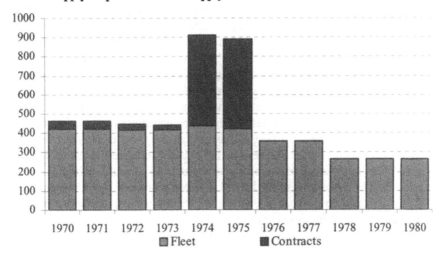

Figure 8.6:   Smedvig's Fleet and Contracts, 1970-1980 ('000 dwt)

Source: Smedvig database.

Smedvig, which in the interwar period owned mainly second-hand bulk carriers, invested primarily in new tankers after 1950. Intending to reap the benefits of the increasing economies of scale in this sector, Smedvig ordered large vessels: MT *Vestalis* was the largest vessel in Scandinavia when it was delivered in 1963, and TT *Veni* was Norway's largest ship when it was launched in 1969. In the early 1970s Smedvig's tanker fleet consisted of a turbine tanker, three motor tankers and a combination carrier, about 400,000 dwt in total. The turbine tanker, the 227,425-dwt TT *Veni*, represented more than half the company's tonnage. This vessel had been delivered in 1969, which means that the company's tonnage doubled from 1969 to 1970. In addition, in 1970 Smedvig had two gas tankers on order at the Moss-Rosenberg yard. The figures in figure 8.6 are slightly misleading because gas tankers, despite their

---

[66]Nerheim and Utne, *Under samme stjerne*, 97.

modest dwt, are relatively costly since they embody advanced technology.[67] In the early 1970s, for example, a 75,000-cubic metre Liquid Petroleum Gas (LPG) vessel was as expensive as a 200,000-dwt oil tanker.[68]

There was no growth in Smedvig's tanker fleet during the 1970s. Due to the sale of the 18,775-dwt MT *Venator*, the size of the fleet was reduced between 1971 and 1972. Subsequently, the tonnage increased three years in a row because of the delivery of supply ships and gas tankers. The company sold MT *Venita* and MT *Vestalis* to Greek shipowners in 1974 and 1975, respectively (see table 8.4). The combination carrier MS *Vestan* transported oil and iron ore until it was sold to Bangladeshi interests in 1977. At the end of the 1970s, TT *Veni* was the only oil tanker left in Smedvig's fleet, and it was disposed of in 1980. Despite this apparent stability, there were in fact large changes in the structure of Smedvig's fleet during the decade. In 1970 the company owned a fleet of more than 400,000 dwt, consisting solely of oil tankers and combination carriers. During the decade the company gradually disposed of its oil tankers while investing in gas tankers and supply ships.

The investments in gas tankers were not always successful economically.[69] LPG/C *Vestri*, the first of the gas tankers, was employed throughout most of the 1970s. After delivery in 1972, it was operated in the spot market until 1976, when it secured a one-year charter. As a result of the high freight rates for gas tankers in 1974, half of the ship's original price was recovered within the first three years of operation.[70] On the other hand, strong competition in the gas transport market caused the second gas tanker, LNG/LPG/C *Venator*, to be laid-up from September 1975 until May 1979.

---

[67]Smedvig's new building contracts, measured in dwt, comprised only ten percent of the company's total fleet in 1972, compared to a corresponding figure for Hilmar Reksten of almost 130 percent. But the amount of money invested in the gas tankers was much larger than an investment in similar-sized oil tankers or bulk carriers would have been. Figures depicting contracts measured in dwt thus underestimate the financial commitment.

[68]See Fearnley and Egers Chartering Co. Ltd., *Review 1977* (Oslo, 1978), table 16, for an introduction to the prices of Liquid Natural Gas (LNG) and Liquid Petroleum Gas (LPG) vessels from 1972 onwards.

[69]In addition to the two fully-owned vessels, Smedvig owned a quarter of LPG/C *Gas Lion*.

[70]Per Gustav Blom, Knut Morten Fjellstad and Frode Nessen, "Investeringsatferd i norske rederier – en casestudie av rederiet Peder Smedvig A/S" (Unpublished thesis, Norwegian School of Economics and Business Administration, 1983), 27; and Nerheim and Utne, *Under samme stjerne*, 224.

Table 8.4
Operation of Four Smedvig Vessels, 1973-1978

	MT *Vestalis* (63,715 dwt)	MT *Venita* (18,000 dwt)	MS *Vestan* (93,026 dwt)	TT *Veni* (227,425 dwt)
1973	Charter until 04/77	Charter until 11/74	Charter until 08/74	Charter until 05/75
1974	Charter, later spot market	Charter until 11/74, then sold	*Vestalis'* old charter from August	Charter until May, then laid-up
1975	Charter until August, sold		Charter until 04/ 1977	Laid-up
1976			Spot from April	Laid-up
1977			Spot until May, sold	Laid-up
1978				Laid-up

Source: See figure 8.6.

Smedvig's decision to operate his gas tankers in the spot market reflected an established strategy. According to Per Blom, Knut Fjellstad and
Frode Nessen, the company's strategy in the 1950s and 1960s can be summarised as follows: "With long-term contracts on existing assets as a stable and
secure source of income, the company would invest in new and risky projects,
which, during a short period of time, could result in losses which would not
affect the company's liquidity and financial strength."[71] The gas tankers represented the risky element; the oil tankers provided the secure revenues. In fact,
Smedvig usually operated these vessels on medium-term charters, which meant
that all his vessels were in secure employment well into 1974 when the tanker
market collapsed. But another consequence was that the firm did not benefit
from the extremely high rates in the spot market before the oil price increase.

On the surface, it might seem that Smedvig stuck to his old business
policy. The reality, however, was not that simple, for in addition to the gas
tankers Smedvig had embarked upon another risky venture. In September
1973, immediately before the oil price increase, he ordered a 472,000-dwt
tanker from a German yard for delivery in 1977. Unlike previous tankers, this
vessel was intended for the spot market. This reflects one of the most important aspects of Norwegian strategy: the potentially fatal effects of the tanker
crisis were not a result of offering the existing fleet in the spot market but
rather in placing orders for new vessels without first securing employment.
The ordering of building number 87 from Howaldtswerke-Deutsche Werft, at a
cost of 230 million German *marks*, sowed the seeds of Smedvig's serious problems. At least three factors made this order problematic. First, the market for
tankers of this size had almost disappeared. Second, since the vessel was a

[71]Blom, Fjellstad and Nessen, "Investeringsatferd i norske rederier," 3.

turbine tanker, it would not necessarily be competitive after the increase in the price of bunkers. Finally, the contract contained an escalation clause that made Smedvig liable to pay more if German wages increased by more than ten percent annually.

Early in 1975, Torolf Smedvig decided to cancel the contract. Unfortunately for him, the Germans were unwilling to accept the cancellation unless Smedvig placed new orders for other types of tonnage. The company, on the other hand, was not interested in new vessels and asked the yard how much a cancellation would cost. Howaldtswerke replied that if Smedvig chose not to order substitute tonnage he would have to pay a cancellation fee of 150 million German *marks*. After negotiations, Smedvig agreed to pay a cancellation fee of approximately NOK 100 million. The result of the arbitration court settlement between Hilmar Reksten and the Aker group became a guideline for bargaining about the size of the fee. Gunnar Nerheim and Bjørn Utne have claimed that "[d]espite the size of the amount, this loss can, in hindsight, be regarded as a good investment in the continuing existence of the company."[72]

Blom, Fjellstad and Nessen have claimed that Smedvig never wanted to have debts of more than fifty percent of its assets.[73] Nonetheless, the firm had problems paying the instalments on its debt in the summer of 1975 because TT *Veni* and the LPG *Vestri* were unemployed, and the prospects for revenue from the drilling rig *West Venture* were bleak due to an oversupply of such vessels. Moreover, the company had to pay parts of the cancellation fee to Howaldtswerke. Despite the sale of MT *Vestalis*, its liquidity was poor.

In contrast to other Norwegian shipowners in similar difficulties, Smedvig chose not to participate in the GI for reasons of principle and ideology. Torolf Smedvig thought that the government should not be burdened with problems that shipping managers had inflicted upon themselves[74] The company preferred instead to negotiate with its creditors for a three-year moratorium on payments. All the creditors, except the French Crédit Lyonnais, accepted a one-year moratorium effective from July 1976. Despite this agreement, however, the company still lacked the freedom of action it wanted.[75]

---

[72]Nerheim and Utne, *Under samme stjerne*, 247, 248 and 250.

[73]Blom, Fjellstad and Nessen, "Investeringsatferd i norske rederier," 31, claim that this saved the company during the shipping crisis. See also Nerheim and Utne, *Under samme stjerne*, 181.

[74]Nerheim and Utne, *Under samme stjerne*, 253.

[75]Due to the agreement with the creditors, Smedvig could not buy an Aker H-3 drilling rig from the GI, despite the fact that the company's management saw this as a profitable investment.

Smedvig is a good example of a shipping company that grounded its strategy in a combination of secure and risky engagements. Since it did not have many vessels laid-up, the immediate effects of the freight rate collapse were small. Instead, its problems stemmed from a single unfortunate decision to contract an unchartered ULCC just before the freight market broke down.

The fate of Rederiet Peder Smedvig illustrates several characteristics of Norwegian shipping. The sale of TT *Veni* in 1980 marked a temporary exit from the tanker sector. In the mid-1980s its gas tankers were transferred to the Singapore registry to reduce operating costs but were taken back on five-year time charters. Smedvig undertook the shift from shipping to the offshore which was characteristic of several Norwegian shipping companies in the wake of the crisis. Today the firm is a leading offshore drilling contractor with a strategy based on medium- and long-term contracts; its fleet includes three mobile rigs, a drill ship and eleven tender rigs operating in international waters.

Sig. Bergesen d.y.

The fate of Sig. Bergesen d.y., Norway's largest shipping company in the 1970s, illustrates the heterogeneity of the Norwegian shipping sector. If the history of Norwegian shipping during the crisis can be characterised by a focus on tankers and bulk carriers, over-eager contracting and unfortunate chartering strategies, particularly with regard to orders for large tankers, the experience of Sig. Bergesen d.y. differs on several accounts. Through advantageous sales of new building contracts, a focus on long-term charters and relatively cautious fleet expansion, the company managed to consolidate its position and in several respects was stronger by the beginning of the 1980s.

This fortunate result was not obvious from its position in early 1973. The company had pooled its resources in the market for large tankers, apparently making it more vulnerable than even Knut Knutsen OAS. Moreover, it was eagerly placing orders for new tonnage and in early 1973 had eleven mammoth tankers and one 223,000-dwt ore/oil carrier on order. But the firm differed from the norm in its chartering strategy and managed to reduce the volume of its new building orders when the tanker market collapsed.

The experience of Sig. Bergesen d.y. illustrates the substantial differences in economic performance between owners with secure revenues from long-term charters and those who fell victim to the depressed spot market. Moreover, it shows how some strategic decisions – in this case to sell four new building contracts – can fundamentally alter a company's economic position. Due to its strong financial standing, Bergesen could use accumulated resources to consolidate its position during the crisis, taking over potentially profitable contracts and engagements from owners in financial distress.

The roots of Sig. Bergesen d.y. can be traced back to the late 1880s, when Sigval Bergesen established himself as one of the most important managing owners in the west coast town of Stavanger.[76] By 1900 Bergesen had become the largest steamship owner in the town. His youngest son, Sig. Bergesen d.y., joined the company in 1916 and became a partner two years later.[77] In subsequent years Sig. Bergesen d.y. assumed responsibility for daily management, while his brother took care of chartering. In 1935 Sig. Bergesen d.y. left his father's firm, spurred by conflicts of interest and personal disagreements.[78] The split was triggered when Sig. Bergesen d.y. contracted tanker tonnage in Denmark without his father's consent. He then spent four years in Denmark, working with the Danish shipowning giant A.P. Møller. By 1940 Sig. Bergesen d.y. had amassed a fleet of three tankers totalling 45,000 grt. During World War II Sig. Bergesen d.y. acquired the yard Rosenberg Verft, which from 1943 until 1970 was an independent division of the Bergesen corporation. In the first postwar decade Bergesen's fleet expanded rapidly; with ships from both Rosenberg and foreign yards, by 1955 he owned a fleet of seven vessels amounting to almost 185,000 dwt.

Four-fifths of the tonnage built at Rosenberg in the 1960s went to Sig. Bergesen d.y.[79] This was supplemented by a considerable number of new buildings from other yards, in particular from Japan. By 1970 Sig. Bergesen d.y. owned eighteen vessels, ten of which had been built in Japan. The fleet amounted to more than 2.2 million dwt, making Sig. Bergesen d.y. Norway's largest shipowner. In 1976 he withdrew from management due to failing health. The other three partners, his grandsons Petter Sundt and Morten Sigval Bergesen, and Jacob Erland Jacobsen managed the company's affairs.[80]

---

[76]For an introduction to the early period, see Jonas Schanche Jonasen, *Sigval Bergesen 1863-1956* (Stavanger, 1963); and Tore Jørgen Hanisch and Liv Jorunn Ramskjær, *Firmaet Sigval Bergesen, Stavanger: under vekslende vilkår 1887-1987* (Stavanger, 1987), 9-98.

[77]Ole Bergesen, three years younger than Sig. Bergesen d.y., became a partner in 1918. For a short biography, see *Norsk Industri*, No. 4 (1975), 17.

[78]See Hanisch and Ramskjær, *Firmaet Sigval Bergesen*. 94.

[79]Ole Felix Dahl, "Med Bergesen som los og kaptein – AS Rosenberg Mekaniske Verksted 1945-70" ["Bergesen as Pilot and Captain – AS Rosenberg Mekaniske Verksted 1945-1970"] (Unpublished hovedoppgave thesis, University of Bergen, 1995), 90; and Hans Mjelva, "Tre storverft i norsk industris finaste stund: ein komparativ studie av Stord Verft, Rosenberg mek. Verksted og Fredrikstad mek. Verksted 1960-1980" ["Three Major Yards in the Finest Hour of Norwegian Manufacturing: A Comparative Analysis of Stord Verft, Rosenberg mek. Verksted and Fredrikstad mek. Verksted 1960-1980"] (Unpublished PhD thesis, University of Bergen, 2005).

[80]Sig. Bergesen d.y. and Co., *Report and Account for 1976* (Stavanger, 1977), 6.

Judging by the structure of its fleet, Sig. Bergesen d.y. would be extremely vulnerable to changes in the tanker market. Like Hilmar Reksten, most of the resources were concentrated in the market for large tankers. Bergesen also owned combination and bulk carriers, and diversified into gas transport towards the end of the decade. Some of the vessels flew foreign flags as a result of cooperation with foreign shipping and mineral consortia.

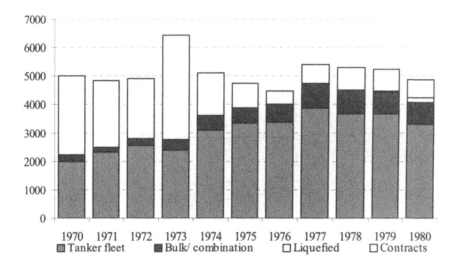

Figure 8.7: Bergesen's Fleet and Contracts, 1970-1980 ('000 dwt)

Note: Two contracts listed in the *Norwegian Shipping News* have been excluded because they relate to the ill-fated combination carriers *Berge Istra* and *Berge Vanga* that were not included in Bergesen's Norwegian fleet. Moreover, only vessels registered in Norway are included.

Source: Bergesen database.

Bergesen's fleet expanded from 2.8 million dwt in 1973 to more than 4.2 million in 1980, after peaking at more than 4.7 million dwt in 1977 (see figure 8.7). The expansion was accomplished through a combination of new building contracts and the purchase of second-hand vessels from shipowners in economic distress. Nerheim has called Bergesen a "courageous Greek" because he recognised the advantages of large tankers before other Norwegians.[81]

---

[81]Gunnar Nerheim, Lars Gaute Jøssang, Bjørn S. Utne and Frida Dahlberg, *I Vekst og Forandring – Rosenberg Verft 100 år 1896-1996* [*Growth and Change – Rosenberg Verft 100 years 1896-1996*] (Stavanger, 1995), 211. Greek shipowners and international oil companies took advantage of economies of scale before most Norwe-

The firm focussed on large vessels because they were "far more inexpensive to operate, and thus counteracted part of the cost development."[82] As a result of this focus, the average size of the vessels in Bergesen's fleet was considerably higher than the national norm (see figure 8.8).

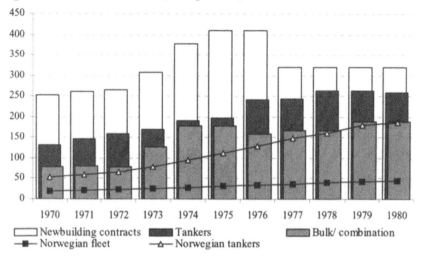

Figure 8.8: Average size, Bergesen's Fleet and Contracts, 1970-1980 ('000 dwt)

Note: Vessels registered abroad, gas tankers and gas tanker contracts are excluded
Source: See figure 8.7

Although its share of tanker tonnage and the average size of its vessels were high, the company was less affected by the crisis than most Norwegian owners. The main reasons for this were its reliance on long-term charters and the fortunate sale of new building contracts prior to the freight market collapse. The strategic aspect that most clearly distinguished Sig. Bergesen d.y. from the other companies was the former. Like Hilmar Reksten, Sig. Bergesen d.y.'s belief in his chartering strategy bordered on the religious. But unlike Reksten, Bergesen was adamant that long-term charters were preferable to the spot market. In 1971 Hilmar Reksten sent Bergesen almost a kilo of documents that contained calculations "proving" that his spot-market strategy was superior to Bergesen's reliance on long-term charters. The reply from the latter was sarcastic: "I have skimmed [your unconventional and most unexpected

---

gians, who did not realise the benefits of large vessels in any great numbers until the early 1970s.

[82]Interview with Sig. Bergesen d.y., *Norsk Industri*, No. 4 (1975), 16.

Easter greeting], but I am searching in vain for its head and tail, as I suppose the documentation should serve another purpose than showing that you are more clever than I am."[83] This was not Reksten's first attempt at converting Bergesen, but the latter had previously "appeared to be disagreeable and immune to approaches that could have hindered some of the harmful effects that the company's chartering policies entailed."[84]

Table 8.5
Fleet Employment, Reksten and Bergesen

Market	Hilmar Reksten				Sig. Bergesen d.y.			
	Mid-1976		June 1979		Mid-1976		June 1979	
	No.	'000 dwt	No.	'000 dwt	No.	'000 dwt	No.	'000 dwt
Period market	(1)	(117)	2	576	20	4073	15	3925
Spot market	2	514	6	1581			7	1041
							(6)	(988)
Laid-up	10	2241	4	892				
Unknown					5	588		

Note: Figures in parentheses refer to tonnage on order. The estimates of the Bergesen fleet from Drewry and the Bergesen database differ because the former includes vessels registered outside Norway but excludes gas tankers.

Source: H.P. Drewry Shipping Consultants Ltd., *The Role of Independent Tanker Owners* (London, 1976), 40 and 43; and Drewry, *Independent Tanker Owners* (London, 1979), 38 and 42.

Reksten's criticism was to some extent justified, as rising costs and unfavourable exchange rates could sometimes make Bergesen's charters unprofitable. For example, in 1976 MT *Bergehus* had six years remaining on a charter that was clearly unprofitable for the company. The charter was cancelled effective in April 1978 after Bergesen paid the charterer US $3.5 million in compensation.[85] Nevertheless, such expenses were relatively small compared with the costs incurred by Reksten as a result of his unemployed tankers. Moreover, Bergesen forfeited the freight rate peaks because of the long charters. But the virtue of this policy was that he was able to secure his

---

[83]Erling Borgen, *Huset Bergesen* [*The House of Bergesen*] (Oslo, 1984), 140-141.

[84]Hilmar Reksten, "Sig. Bergesen d.y. and Co. 1947-1970" (unpublished manuscript sent to Sig. Bergesen d.y. and Norwegian newspapers, 1971), 73.

[85]Bergesen, *Report and Account for 1976*, 4.

income during the troughs in the spot market. Two analyses from the British shipping analysts H.P. Drewry illustrate the differences in the position of Reksten and Bergesen after the freight market collapse. While Reksten had been forced to lay-up the majority of his tonnage, Bergesen's fleet was assured revenues in 1976 (see table 8.5).

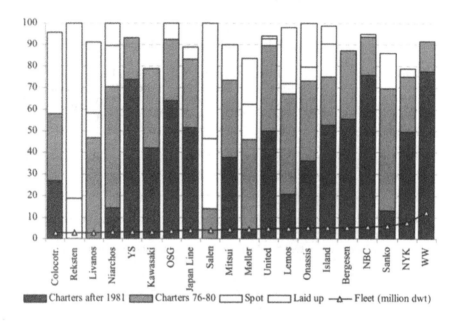

Figure 8.9   Exposure of the World's Twenty Largest Independent Tanker Owners, Mid-1976

Note: The bars show the manner in which the companies' vessels were employed. Vessels for which charter expiry dates or employment were unknown are excluded, thus giving aggregate percentages of less than 100 for several companies. The line shows the aggregate fleets of the companies in million dwt.

Source: Drewry, *Role of Independent Tanker Owners*.

In mid-1976 approximately eighty percent of Bergesen's fleet was on charters that expired in 1978 or later, and more than fifty-five percent had secure employment after 1981. In an international perspective, the share of the Bergesen fleet with secure employment was high. Indeed, only two of the largest tanker owners in 1976 were in a more secure position. The least exposed company was Sir Yue-Kong Pao's World Wide group, by far the largest tanker owner with a fleet of more than twelve million dwt. A major proponent of the *shikumisen* agreements, Pao had tied more than ninety percent of his fleet to Japanese charterers. But the economic problems of some of the charterers, notably the Japan Line, could have spelled disaster.[86] The other company with a large degree of charter cover was D.K. Ludwig's National Bulk Carriers, which had secure employment after 1981 for seventy-six percent of its fleet.[87] Figure 8.9 indicates the extent to which the world's twenty largest independent tanker owners were exposed to the difficulties of the mid-1970s. The companies have been ranked by the size of their fleets. The figure illustrates the share of the companies' tonnage on charters expiring in the period 1976-1980; the share of tonnage on charters after 1981; and the share of tonnage laid-up or operating in the spot market.

The difference between the two Norwegian participants is striking. Whereas Reksten had been forced to lay-up ten of his twelve vessels, Bergesen had no ships laid-up on his own account, and at least eleven of his vessels were assured employment well into the 1980s. Moreover, due to structural changes in the chartering market it became increasingly difficult for Bergesen to secure long-term employment when the current charters terminated. Accordingly, approximately a fifth of the fleet traded in the spot market in 1979.

Bergesen was the world's fifth largest independent tanker owner in 1976. Three years later the company had been relegated to eighth place, but it was still fifth if tonnage on order is included.[88] One reason for the "decline" was that Bergesen, in contrast to eighty percent of the other large independents, had no backlog of new building orders in mid-1976.[89] Indeed, although expansive in absolute terms, Bergesen's contracting immediately prior to the freight market collapse can be considered cautious. Yet judging by tonnage on order in early 1973, Bergesen's contracting policy seems anything but conservative. The company had twelve vessels above 200,000 dwt on order, meaning that the volume of contracts was approximately one-third higher than its existing fleet. But when the tanker market collapsed that autumn, four of the con-

---

[86]See Stokes, *Ship Finance*, 59-60. Since Pao had secured the income from his charterers through bank guarantees, he was less exposed than shipowners who had signed charter agreements with other companies in financial distress.

[87]The Japanese Yamashita-Shinnihon Steamship Co. and the New York-based Overseas Shipholding Group Ltd. also had higher shares of their fleets on charter into the 1980s. But both were smaller than Bergesen, and the amount of tonnage on charter was thus lower.

[88]Drewry, *Independent Tanker Owners*, 4. Chartered-in vessels have not been included in any of the analyses.

[89]The tonnage on order in 1976 in figure 8.8 refers to *Berge Empress*, which was delivered in April 1976.

tracts had been sold. As a result, Bergesen's new buildings on order in January 1974 corresponded to forty percent of the existing fleet, rather than the 133 percent it had been a year earlier. Its order book was thus low, compared with 106 percent of the existing fleet for Norwegian and eighty-two percent for international shipowners.[90]

Sig. Bergesen d.y.'s ordering at times was radically different from other Norwegian shipowners. In the summer of 1973, shortly before the tanker market collapsed, the company offloaded two 400,000-dwt contracts at Kawasaki to Italians, pocketing US $40 million in the bargain. The sale of these contracts was prompted by a fear of overcapacity and the fact that employment had not been secured, contrary to the company's normal policy. Earlier the same year, Sig. Bergesen d.y. had sold two contracts in Denmark to the Greek shipowner Livanos at a profit of approximately US $30 million.[91]

In 1977 Sig. Bergesen d.y. "caused a small commotion – at home and abroad" when in the midst of the tanker gloom it ordered two 320,000-dwt tankers in Japan.[92] This order, which allegedly cost in the region of NOK 500 million, was one of several expansionary moves made by the company in the depressed tanker market. While the idea of new building contracts in a market characterised by massive overcapacity may seem irrational, the vessels had diesel engines, which made them advantageous compared with the turbine-driven tankers available on the second-hand market. Moreover, the new buildings were equipped with segregated ballast tanks in anticipation of changes in international maritime legislation. Still, the contracts were controversial. Some in the industry saw them as an indication that vessels supported by guarantees from the GI might not be worth keeping. In an internal memorandum the Norwegian Shipowners' Association claimed that "[t]he price indicated is approximately NOK 500 million for the two vessels. However, we know from other sources that the price was particularly favourable, due to factors that can not be disclosed."[93] The memorandum speculated that the order might have been connected to particular aspects within the company rather than to the state

---

[90]The order book as a share of the fleet refers to tankers and combination carriers and has been calculated based upon Fearnley and Egers Chartering Co. Ltd., *Review 1974* (Oslo, 1975), tables 3, 9, 20 and 26.

[91]Borgen, *Huset Bergesen*, 139-144.

[92]Quote from *Norwegian Shipping News*, No. 8B (1977), 34.

[93]Quote from Archives of the Norwegian Shipowners' Association, folder 6 B K 75 – Krisen 1975 VII, 010177-300977, "The Tanker Contracting of Sig. Bergesen d.y. and Co.," 2 May 1977.

of the tanker sector in general. One possible explanation suggested by the author was incentives with regard to taxation.

Some of Bergesen's new building contracts in the late 1970s were a direct result of the crisis and its effects on other Norwegian shipowners. Several of the gas tankers had originally been ordered by the Norwegian shipowners Fearnley and Eger, but due to considerable economic difficulties the company was unable to honour them. Sig. Bergesen d.y. took over the contracts, signing a financing agreement with the yard in the process. The company's entry into the gas sector – which was to be the main focus in the 1980s and 1990s – was thus partly accidental. The type of vessels ordered by Bergesen did not deviate significantly from other Norwegian owners – they ordered the largest and technologically most advanced tankers and combination carriers available. But in some ways Bergesen can be seen as "first among equals" because its focus on the most efficient tonnage was a way to neutralise the potentially harmful effects of long-term charters. Moreover, through the sale of four new building contracts in 1973, Bergesen managed to avoid an increase in the fleet which would have proved unfortunate.

Hilmar Reksten's new building orders were so large that adaptation to the changed market was virtually impossible. If Bergesen had maintained the high volume of tonnage on order, his situation would have been less precarious, as long-term charters secured the revenue from the company's existing fleet. Nevertheless, the new buildings, and the fact that they did not have charters, would have represented a considerable drain on the company's resources. It is thus evident that whereas the chartering strategy was paramount to the company's relative success in the 1970s, the sale of the new building orders was also important in explaining its favourable position.

Bergesen's financial strength was reflected in the fact that the company maintained large bank deposits throughout the crisis. Shortly after the tanker market collapsed, the company's liquid reserves amounted to some NOK 700 million.[94] Two years later its bank deposits had increased, and Sig. Bergesen d.y. reported the highest operating surplus in its history – in a period when most others in the shipping sector were concerned with the tanker crisis.[95] The company's liquidity continued to improve, in stark contrast to the majority of Norwegian shipowners, for whom the crisis represented a drain on accumulated reserves.[96]

---

[94]*Norsk Industri*, No. 4 (1975), 17.

[95]In 1976 the result was NOK 362 million, compared with NOK 321 million in 1975; see Sig. Bergesen d.y. and Co., *Report and Account for 1976*.

[96]By 1979 liquidity had increased to NOK 839 million; see Sig. Bergesen d.y. and Co., *Half Year Report for the Shareholders in Companies Managed by Sig. Bergesen d.y. and Co.* (Stavanger, 1979), 3.

The advantageous financial situation was a result of the fact that the company's vessels were earning revenues rather than being laid-up on Bergesen's account. In essence, profitability was not affected by the state of the tanker market at any given time. Rather, the important factor was the condition of the market at the point at which the charters were signed.[97] At the same time, the company's debt servicing was relatively modest, partly because it had managed to offload some of the new building contracts. Despite the focus on the most depressed part of the shipping market, the prudent chartering strategy ensured that there was no need for Sig. Bergesen d.y. to approach the GI for assistance. But the company managed to take advantage of the vessels sold by other shipowners who had been forced to seek assistance from the GI.

In late 1981 Bergesen established a partnership, of which it owned fifty-one percent, with the Biørnstad group, controlled by the GI, which held the remainder. The co-owners took over three vessels previously owned by the ill-fated Biørnstad companies, and Bergesen became responsible for managing and chartering them.[98] The following year Bergesen again went into co-ownership with the GI in connection with the former Reksten tankers TT *Cyprian* and TT *Julian*, and also managed the ex-Reksten tanker TT *Trajan*, owned by the GI.[99]

Contrary to the majority of Norwegian shipping companies, Sig. Bergesen d.y. was actually strengthened by the shipping crisis. In the mid-1980s the family's private ownership ended as Bergesen d.y. ASA became a public company listed on the Oslo bourse. In 2003 it was taken over by the Sohmen Pao family, owners of the World Wide tanker fleet. The new owners took the company off the stock exchange and consolidated the large tankers into Bergesen Worldwide Ltd. In October 2005 Bergesen Worldwide Gas, the

---

[97]The collapse implied that when the vessels were re-delivered to Bergesen they would be affected by the depressed market. But due to the focus on long-term charters the crisis could carry on for a number of years before Bergesen would have been influenced.

[98]Norway, Parliament, Stortingsproposisjon No. 38 (1981-1982), "Om bevilgning til utbetaling under garanti stillet av Norsk Garantiinstitutt for skip og borefartøyer" ["On the Appropriation of Pay-out with Regard to Guarantees granted by the Norwegian Guarantee Institute for Ships and Drilling Vessels Ltd."].

[99]Norway, Parliament, Stortingsmelding No. 53 (1981-1982), "Om avtale av 5. februar 1982 mellom Norsk garantiinstitutt for skip og borefartøyer AS, Hambros Bank Limited, R/A Hadrian og R/A Trajan i forbindelse med avviklingen av garantiengasjementene i Reksten-flåten" ["On the Agreement dated 5 February 1982 between the Norwegian Guarantee Institute for Ships and Drilling Vessels Ltd., Hambros Bank Limited, R/A Hadrian and R/A Trajan on the Winding-up of the Guarantee Engagements in the Reksten Fleet"], 8.

world's largest gas carrier company with a fleet of more than eighty ships, was reintroduced to the Oslo Stock Exchange.

Conclusion

The analysis in this chapter has illustrated the differing degrees to which Norwegian shipowners were affected by the shipping crisis, and the four cases discusssed underscore the heterogeneity of the Norwegian shipping sector. Three strategic elements have been emphasised – fleet structure, chartering strategy and contracting and financing. While all the policy parameters contributed to an explanation of the fates of the different companies, chartering and new building strategies appear to have been particularly important. Fortunate decisions in these areas enabled Sig. Bergesen d.y. to emerge stronger from the crisis. Moreover, unfortunate decisions with regard to contracting and chartering – and in particular the combination of the two in terms of unchartered new building contracts – can explain the difficulties of Knutsen and Smedvig, as well as the collapse of Hilmar Reksten's empire.

The fleet structure was important in explaining the unfortunate fate of Norwegian shipowners in general, compared with their colleagues in other countries. But the analysis of individual companies suggests that fleet structure *per se* was not a good indicator of their economic well-being. Rather, the existing fleet structure must be seen in relation to strategic determinants such as chartering and contracting. Judging by fleet structure alone, Knut Knutsen OAS seemed better equipped to cope with the crisis than Sig. Bergesen d.y., but the history of the companies shows that this was not the case.

New building strategies can contribute to an explanation of the misfortune of three of the companies. For Knutsen and Smedvig, orders signed shortly before the freight market collapse represented a considerable drain on their resources. But if the new building contracts had been secured through profitable long-term charters, the situation would have been less precarious. In many ways, the economic difficulties of Knutsen and Smedvig mirror those of several other Norwegian shipowners – the basis for the problems was neither contracting nor chartering strategy *per se*, but rather the combination of these elements, i.e., the ordering of new buildings for which charters had not been secured.

The fate of Hilmar Reksten illustrates the vulnerability of a shipowner who was willing to gamble. Reksten had pooled his resources into that part of the industry where the crisis was most pronounced. Moreover, the sheer volume of contracts would have been sufficient to topple the company when freight rates collapsed. In addition, even in the absence of a large number of contracts, Reksten's chances of sufficiently profitable employment after the collapse would have been minuscule.

Knutsen and Smedvig were both relatively small companies, for which tonnage on order represented a dramatic increase in the size of their fleets. The large amount of resources committed through a single decision, such as the signing of a contract for a large vessel, implied that the companies' fate hinged on the development of certain key markets. The competitiveness of Norwegian shipowners was closely related to the operation of technologically-advanced vessels that could utilise the economies of scale inherent in vessel size. Due to the increase in the average size of ships, individual strategic decisions tied up a disproportionate share of the companies' resources.

The four companies surveyed illustrate some of the changes in the Norwegian shipping sector following the shipping crisis. In the case of Knut Knutsen OAS, the fleet became the foundation for a shipping enterprise managed by new owners. Hilmar Rekstens Rederi became one of the most manifest victims of the shipping crisis. Rederiet Peder Smedvig made a successful transition to the offshore sector, while the fate of Sig. Bergesen d.y. illustrates that there was a continued basis for Norwegian shipowners even after the crisis.

Chapter 9
Conclusion

The crisis of the 1970s and 1980s fundamentally changed the shipping market. A large number of companies were forced to lay-up their vessels; the relationship with creditors became increasingly problematic; and profits diminished. For many, the result was liquidation or bankruptcy. The shipbuilding industry found it ever more difficult to attract new orders. Although subsidies provided a temporary reprieve, from the late 1970s onwards the industry was downscaled and lost its historic position, especially in Western Europe. Several traditional maritime nations were unable to adapt to the new conditions and suffered from a massive disinvestment in shipping.

This book has had two aims. The first was to analyse the causes and effects of the most serious shipping crisis of the twentieth century. The second was to explain why Norwegian shipowners were more severely affected than their foreign competitors, and to trace the effects of the crisis on the Norwegian shipping sector.

The situation in the early 1970s represented a dramatic break with the first postwar decades. When this "golden age" ended, most sectors had to readjust from rapid expansion to lower growth. But the transformation in shipping was more extensive than for many other sectors. There were a number of reasons for this. The pre-crisis development had been particularly positive, while the crisis was both more severe and prolonged than in other sectors. Moreover, the shipping industry cannot "smooth" demand fluctuations by producing for stock. After 1974 demand dwindled while the fleet increased, resulting in a severe imbalance. In the first part of the 1980s an absolute reduction in oil transport aggravated the already difficult state of the market.

On the demand side, two developments were particularly important. First, the oil price increases initiated by the Organisation of Petroleum Exporting Countries transferred the status of oil from a source of energy to an instrument of economic and political warfare. Oil transports – the driving force behind shipping expansion in the 1950s and 1960s – could no longer be counted upon to produce annual two-digit growth rates. Second, the world economy underwent a transition from high growth to recession and limited trade expansion. Reduced activity affected trade and transport demand for some of the most important bulk commodities. To make matters worse, the demise of the gold standard based upon the US dollar led to substantial exchange rate fluctuations, leading to uncertainty and guaranteeing greater risk for shipowners.

With regard to supply, the crisis was exacerbated by the high volume of new building contracts signed before the crisis. In this respect, two elements that have to be considered are the basis for the widespread ordering before the collapse and growth of the world fleet continued even after the massive over-supply was evident.

In early 1974, the volume of tankers and combination carriers on order represented more than eighty percent of the existing fleet. The basis for these large orders can be understood by looking at three groups: shipowners, shipyards and financial institutions. Shipowners and oil companies expected that demand for oil transport would continue to grow and positioned themselves accordingly. In particular, the high freight rates in 1973 induced many owners to order new tonnage. The massive subsidisation of the shipbuilding industry reduced the "real" cost of increasing transport capacity, thus inflating fleet growth. The access to inexpensive financing enabled companies with limited financial resources to increase their gearing, which meant that some of the risk of ordering new tonnage was transferred from the owners to their creditors. The combination of low real interest rates and beneficial tax deductions led to negative post-tax real interest rates on shipping investments. The high pre-crisis contracting can thus be explained by the fact that ordering new tonnage was rational given shipowners' expectations about future developments, which in turn were strongly influenced by the historical record and the influence of banks and governments. The profitability of individual investments was overestimated due to the strong competition among shipyards and bankers and the influence of tax regimes.

The lag between contracting and delivery, which is particularly long in periods when there are large order books, was the main reason for the continuing growth of the tanker fleet after the freight market collapse. Some of the vessels on order in 1973 were scheduled for delivery in 1978/1979. Another reason for the relatively slow adaptation to the oversupply was that the maritime sector was unable to fathom fully the extent of the crisis. Rather, it viewed it as a temporary phenomenon of the kind often experienced in shipping. Cyclical downturns are a regular feature and are – at least in the short term – difficult to differentiate from the kind of fundamental shifts experienced in the 1970s and 1980s. The limited cancellations indicate that shipowners did not recognise that the crisis was so severe that paying cancellation fees was preferable to accepting delivery.

The tanker supply surplus soon spread to other parts of the shipping market. This was done initially by combination carriers previously operating in the tanker segment. The conversion of new building contracts increased the non-tanker fleet in the medium term. In the long term, shipyards offered generous terms to attract orders. Moreover, falling demand following the economic recession added to the problems in the non-tanker sector.

In short, the shipping crisis was the result of a tremendous disparity between expected and actual events. Supply increased, based upon expectations of strong growth. Due to the oil price increase and the recession, actual demand growth was negligible or even negative for some segments. By the time people realised that the expected growth would not materialise, it had become impossible to adapt supply to the new conditions.

The first effect of the disruption was a drastic fall in tanker freight rates and reduced activity in the charter market. Some rates fell by more than eighty percent within a month. Tanker lay-ups, which were negligible in 1973, increased to more than eighteen percent of the fleet in the spring of 1976. Reduced average speeds became more profitable as bunker prices increased, and this contributed to a reduction of the surplus. The use of tankers for storage and part cargoes, as well as extra waiting time in ports, also reduced lay-ups. The tanker fleet continued to grow until 1978. Owners were hesitant to scrap underutilised assets because they hoped that a market recovery would increase the value of their tonnage. Due to the strong correlation between freight rates and second-hand prices, the value of the world fleet plummeted.

The shipping crisis led to serious liquidity problems for many shipping companies, particularly those with large tanker investments. The value of a standard 200,000-deadweight ton (dwt) tanker fell by eighty percent from 1973 to 1975. As profit margins were squeezed, substantial debt burdens and high costs forced owners in Traditional Maritime Nations (TMNs) to reduce operations or to transfer their tonnage to flags of convenience (FOCs). The large disinvestment in shipping in the Organisation for Economic Co-operation and Development countries was reflected in the increasing importance of Emerging Maritime Nations (EMNs), particularly in Asia.

The supply growth is particularly important in explaining the predicament of the shipping industry in the 1970s. The transport capacity of the world fleet increased by more than half from 1973 to 1977, spurred by an increase in the tanker fleet of more than two-thirds. This increase exacerbated the negative effects of the reduced demand. Demand developments were more important for the increasing imbalance in the 1980s – the second stage of the shipping crisis. The transport of crude oil fell by more than sixty percent in the period 1977-1985, contributing to a twenty-five percent reduction of seaborne transport. Even a decrease in the world fleet was insufficient to neutralise the effects.

Norwegian shipowners had been particularly successful during the growth period of the 1950s and 1960s. They were eager in the new building market, ordering new tonnage that exploited economies of scale and found profitable employment. Even though there were periods of slack demand, profits during freight rate peaks ensured that overall returns were positive. The most beneficial investments were large, fast vessels, especially tankers. Many Norwegian owners channelled a large portion of their resources into the fast-growing bulk sector. A large share of the fleet consisted of mammoth tankers and flexible combination carriers, which allowed owners to reap the benefits of favourable demand before the oil price increases.

The oil price increases led to structural changes in the markets for shipping services and energy. The unfavourable demand situation meant that the tables had been turned. Those who had benefited the most from developments in the 1960s were the ones least able to adapt to the new circumstances. The vessels that had previously been among the most advantageous became particularly ill-suited to the new conditions. The state of the shipping market in the 1970s was diametrically opposed to the situation in the 1960s. As demand growth faltered and bunker prices increased, a fleet consisting of large turbine tankers became a burden rather than an advantage. Once the market was characterised by oversupply, a large "fleet" of new building contracts became a liability rather than an asset. The winners of the 1960s became the losers of the 1970s.

Norwegian lay-ups were considerably higher than the international average, and Sweden was the only country with lay-up rates at the same level. The flight from the Norwegian flag – vessels sold to foreigners or re-flagged to low-cost countries – was also large in international terms. Moreover, the reduction of the Norwegian fleet occurred earlier and was more pronounced than for any other major shipping nation except Great Britain.

The high lay-up rates and strong reduction of the fleet in the wake of the crisis indicate that Norwegian shipowners were hit harder by the changed shipping market than their competitors. The reasons for the unfortunate Norwegian experience can be found through an analysis of Norwegian owners' strategic choices. Three "policy instruments" – fleet structure, chartering and contracting – played significant roles.

A relatively large share of the Norwegian fleet consisted of vessels operating in the hardest hit segments. In 1975 tankers and combination carriers comprised seventy percent of the Norwegian fleet, compared with an average of less than sixty percent for the rest of the world. Moreover, the average Norwegian ship was more than twice as large as the average international vessel. This implies that the decline in the value of the Norwegian fleet was particularly sharp. In the period following the crisis, Norwegian shipping lost its important position both domestically and internationally. By 1987, the Norwegian fleet had been reduced by almost thirty-eight million dwt, or more than three-quarters, compared with the peak ten years earlier.

Norwegian owners traditionally operated a relatively large share of their tonnage in the spot market. Yet when the market collapsed, their chartering strategy was not much different from that of their competitors. Moreover, a comparison of the return from the various market segments shows that the additional risk in the spot market had been more than sufficiently rewarded through higher average returns before the market collapsed.

Two aspects of chartering policy put Norwegian owners at risk. First, they generally operated on shorter charters than their competitors. Accordingly, they were able to weather a temporary downturn but not a lengthy crisis. Second, they were responsible for a large share of the mammoth tanker tonnage on order. The degree of charter cover for these vessels was very low: more than eighty percent of the vessels on order were intended for the spot market, compared with roughly a third for the international fleet. Hence, it was the shorter duration of charters and the large amount of unfixed tonnage on order that put Norwegian shipowners in a difficult position.

Norwegian owners traditionally had a large amount of tonnage on order. This can be explained by their focus on new, technologically advanced, large vessels. The contracts could either be cancelled, involving expensive cancellation fees, or the new vessels could enter a market where there was little hope of profitable employment. Several large tankers went straight from the building berth into lay-up. For some companies the fees paid to cancel new buildings represented a smaller drain on resources than accepting delivery of a tanker for which no market existed. The high level of orders before the collapse hampered several owners' ability to adapt to changing market conditions. Due to the backlog of ordered but undelivered vessels, the Norwegian fleet continued to grow until 1977, although contracting all but dried up after 1973.

These strategic choices are important for understanding the predicament of the Norwegian shipping industry. But the common view that Norwegian owners were victims of their own willingness to take risks requires modification. More properly, the Norwegian misfortune can to a large extent be understood in terms of a combination of risk propensity and rational decisions based upon business conditions and expectations that did not eventuate.

First, the failure to anticipate the massive changes in the shipping market and the international economy was not the sole province of Norwegian shipowners. Most agents were taken by surprise by both the magnitude of the changes and the length of the crisis. The adaptation to the changed conditions was based on the notion that the downturn was temporary.

Second, the focus on large, modern vessels was mainly the result of the competitive advantage of Norwegian owners in the management of technically advanced tonnage and can be explained by the relative prices of labour and capital. Indeed, Norwegian owners channelled their resources into those segments of the shipping market where they were able to make a profit. It was impossible to foresee that these were the parts where the collapse would be most dramatic. Moreover, the crisis turned out to be so severe and long-lasting that even shipowners who had chosen what was traditionally considered a relatively prudent chartering strategy, with an emphasis on medium-term charters, were severely affected.

Due to the importance of the shipping sector to the Norwegian economy, the crisis prompted public action. The Norwegian government responded in two ways. Initially, it established the Norwegian Guarantee Institute for Ships and Drilling Vessels Ltd. (GI) to assist owners and reduce the tonnage flight. In the early 1980s, restrictions on access to registration abroad were eased, making it possible to take advantage of less expensive foreign crews. The combination of a relatively strict flag policy and the existence of the GI created a pent-up demand for foreign registration. An exodus of tonnage commenced. The Norwegian fleet was reduced by more than seventy-five percent from the late 1970s to the mid-1980s. In the late 1980s, after a reorientation of shipping policy led to the introduction of the Norwegian International Ship Register, the fleet began to grow again.

An analysis of Norwegian shipping companies illustrates the drastic reduction of shipping activity in the wake of the crisis. The tonnage on Norwegian registry fell to less than ten million dwt in 1987, after peaking at almost fifty million dwt ten years earlier. The number of Norwegian ports with companies owning vessels of more than 5000 gross registered tons almost halved between 1970 and 1987. Moreover, the number of owners with ships above this size on Norwegian registry fell by more than 100, representing a decline of more than two-thirds. The number of shipping companies declined by almost half even when those transferring their vessels to foreign flags are included. The crisis coincided with a trend towards increased concentration in the Norwegian shipping industry. This pattern was evident both at the geographic and company levels, as the crisis sounded the death knell for a large number of Norwegian ports and shipping companies with long maritime histories.

Parallel with the reduction of the fleet from 1978 onwards, the Norwegian merchant marine became increasingly diversified, as owners focused on more specialised shipping. A comparison based upon compensated gross register tons, taking into account the sophistication of the vessels, reveals that the reduction of the fleet was far less dramatic than analyses using dwt indicate. This can be explained by the fact that Norwegian owners scaled down their involvement in the bulk sector, focussing instead on more advanced types of tonnage.

The differences in the fate of the four shipping companies considered in this book correspond well with the heterogeneity of the Norwegian shipping industry. Some companies survived relatively unscathed due to advantageous business decisions prior to or shortly after the freight market collapse. For others, the obligations were so large that adaptation was virtually impossible.

The focus on Norwegian owners was partly hampered by the lack of similar analyses of shipowners in other nations. While the Norwegian experience has been contrasted with that of foreign shipowners, studies of a more national character will be important in explaining the role of the various strategic choices. How did the shipping crisis affect owners with strategies different than the Norwegians? For instance, were long charters alone a sufficient basis for Hong Kong shipowners to acquire their important position in the industry?

Parallel with the shipping crisis, maritime hegemony shifted, with an increasing share of the world fleet registered in EMNs in Asia and FOCs. This development continued after the market recovered, although new institutional arrangements in some TMNs have stemmed the outflow of tonnage. Perhaps the most successful manoeuvre was the establishment of the Norwegian International Ship Register, where FOC-like cost advantages were made possible under the domestic flag. Although Norway lost more tonnage than other countries during the crisis, the fleet increased substantially after the market recovered and the new institutional arrangements were in place.

The shipping crisis lasted throughout most of the 1980s, and many of the vessels ordered during the new building spree of the early 1980s continued to criss-cross the world's oceans into the new millennium. Thirty years after the crisis first erupted, the Norwegian shipping analyst Fearnresearch concluded that "2004 was the best year since 1973 for tankers and the best ever for most other vessels types."[1] But several aspects emphasised in the article have a familiar ring:

> Rates and values soared to almost unbelievable levels for new and even very old ships...Rate levels saw substantial downward corrections, most for the largest tankers which saw spot returns tumble about 80% from extreme levels in just a few weeks – and then to levels which are still more than healthy...The very high order book for container vessels is expected to have negative impact on the tonnage balance.

There is little doubt that the shipping industry is highly cyclical. But the hangover from the party of the late 1960s and early 1970s proved both particularly violent and long-lasting. According to Martin Stopford, one of the world's leading shipping analysts, the good times in 2004 and 2005 were influenced by the fact that "the industry has finally shed the crushing burden of the huge oversupply there had been since the 1970s."[2] Whether the agents in the shipping industry have also shed their knowledge about the dangerous effects of demand changes and the lack of moderation is another question.

---

[1] Fearnresearch, "Global Transportation Aspects" (Unpublished mss., January 2005).

[2] *The Economist*, 18 August 2005.

Appendices

Appendix 1
Norwegian Tankers Laid-up, December 1975

Vessel	Shipowner	Dwt	Date	Vessel	Shipowner	Dwt	Date
MT *Acina**	C.H. Sørensen	130	05	TT *Kong Haakon*	H. Reksten	219	10
TT *Aurelian*	H. Reksten	219	02	MT *Kristina*	Grieg and Co.	62	02
MT *Balla B.*	Brøvig	21	03	MT *Lina C.**	Fekete and Co.	20	06
TT *Beaumaris*	Biørnstad	219	06	MT *Long Phoenix*	Lange	51	02
TT *Beaurivage*	Biørnstad	285	05	MT *Molda*	Mowinckel	144	10
MT *Beauval*	Biørnstad	88	07	TT *Mosqueen*	Mosvold	71	07
TT *Belfri*	Belships	311	03	TT *Octavian*	H. Reksten	220	01
MT *Bente B.*	Brøvig	19	02	MT *Orator*	O.B. Sørensen	135	04
MT *Berge Pr.**	Bergesen d.y.	280	04	MT *Osco Spirit*	Schrøder	33	03
MT *Bergevik*	Bergesen d.y.	100	03	MT *Pepita*	Foyn Bruun	20	02
MT *Bjørkaas*	Mørland	18	02	MT *Perikum*	Foyn Bruun	36	08
MT *Britta*	Mathiesen	138	01	MT *Petunia*	Foyn Bruun	26	06
MT *Cerno*	J.P. Jensen	103	06	MT *Polartank**	Melsom	135	04
MT *Cis Brøvig*	Brøvig	106	06	TT *Radny*	Waage	286	11
MT *Condo*	Wangen	77	10	TT *Raila*	Waage	219	02
TT *Corona*	Wrangell	232	11	TT *Ranja*	Waage	219	02
MT *Credo*	Wangen	101	05	MT *Ronaville*	Saanum	36	05
MT *Curro*	J.P. Jensen	134	03	TT *S Ugelstad*	S. Ugelstad	34	03
TT *Daghild*	J.P. Pedersen	256	03	MT *Sea Breeze*	Bendixen	136	07
TT *Elisabeth K.*	Knutsen	216	01	TT *Sir Charles H.*	H. Reksten	286	*12*
TT *Fabian*	H. Reksten	286	01	TT *Sir Winston C.*	H. Reksten	95	12
MT *Fagerfjell**	Olsen and U.	126	*12*	MT *Skaugum*	Skaugen	77	05
TT *Ferncourt**	F&E	245	08	TT *Solfonn*	Sig. Bergesen	132	10
MT *Fernmanor*	F&E	78	03	MT *Solviken*	Wallem/Ste.	21	03
TT *Fernmount**	F&E	216	04	MT *Sunares*	Hansen-Tangen	20	05
MT *GC Brøvig*	Brøvig	17	03	MT *Sydhav*	Lodding	139	04
MT *Gimle*	Evensen	59	03	TT *Synia*	Waage	226	09
MT *Granheim*	Bjørge	21	03	MT *Tamarita*	J.M. Ugland	75	05
MT *Gunvor B.*	Brøvig	19	03	MT *Tank Rex*	Herlofson	52	03
MT *Gylfe*	Evensen	20	01	MT *Templar*	W. Wilhelmsen	83	*12*
MT *HM Wrang.*	Wrangell	87	11	MT *Thorshov*	Thor Dahl	103	08

Vessel	Shipowner	Dwt	Date	Vessel	Shipowner	Dwt	Date
TT *Hadrian*	H. Reksten	220	10	MT *Thor-shøvdi*	Thor Dahl	103	05
TT *Harry B.*	H. Borthen	225	05	TT *Torill Knudsen*	Knutsen	285	06
MT *Havbør*	P. Meyer	44	04	TT *Troma*	Mowinckel	86	02
TT *Hitra*	Mowinckel	38	08	TT *Veni*	Smedvig	227	05
TT *Jagranda*	Jahre	90	06	TT *Vespasian*	H. Reksten	285	02
TT *Jorek Trader*	J. Reksten	99	04	MT *Vinga*	Mowinckel	138	08
TT *Julian*	H. Reksten	286	01	MT *Wangli*	Wangen	126	06
MT *JJ Lorentzen*	Stove Shipping	132	03	MT *Wangskog*	Wangen	126	10
MT *Kollbris**	Bjørge	136	09	MT *Wangstar*	Wangen	97	07
MT *Kollskegg*	Bjørge	136	03	MT *Wilstar*	A. Wilhelm-sen	133	03

Notes: Date refers to the month in 1975 when the vessel was laid-up, except for figures in italics, which refer to 1974. An asterisk indicates a vessel laid-up on
the charterer's account. In addition to the vessels above, three gas tankers,
thirteen combination carriers, four bulk carriers and one passenger vessel had
been mothballed by December 1975.

Source: *Norges Handels og Sjøfartstidende*, 3 December 1975.

Appendix 2
Norwegian New Building Contracts above 200,000 dwt, 1 January 1974

Ship-owner	Dwt		Yard	Country	Delivery	Chartering
Anders Jahre	255	TT	Nippon Kokan	Japan	March 1974	5 year/1979
Anders Jahre	255	TT	Nippon Kokan	Japan	August 1974	5 year/1979
Anders Jahre	267	TT	Nippon Kokan	Japan	May 1977	Unfixed
Anders Jahre	285	TT	Nippon Kokan	Japan	March 1976	Unfixed
A. Jahre (Kosmos)	258	TT	Nippon Kokan	Japan	December 1975	Unfixed
Arth. H. Mathiesen	230	TT	Mitsui Shipbuilding	Japan	1977	Unfixed
Biørn Biørnstad	320	TT	Aker Group	Norway	4 qtr 1974	Unfixed
Biørn Biørnstad	258	TT	Aker Group	Norway	3 qtr 1975	Unfixed
Biørn Biørnstad and Co.	230	TT	Aker Group	Norway	1 qtr 1976	Unfixed
Biørn Biørnstad and Co.	280	TT	Aker Group	Norway	2 qtr 1976	Unfixed
CH Sørensen and Sønner	410	TT	Bremer Vulkan	Germany	Medio 1975	5 year/1980
Einar Rasmussen	410	TT	Mitsui Shipbuilding	Japan	October 1974	4 year/1978
Einar Rasmussen	233	TT	Uddevallavarvet	Sweden	2nd half 1976	Unfixed
Einar Rasmussen	472	TT	Uddevallavarvet	Sweden	Mid 1977	Unfixed
Erling H. Samuelsen	472	TT	Kockums Mekaniska	Sweden	June 1976	Unfixed
Erling H. Samuelsen	236	TT	Kockums Mekaniska	Sweden	December 1977	Unfixed
H Staubo/JP Pedersen	472	TT	Kockums Mekaniska	Sweden	August 1974	3 year/1977

Ship-owner	Dwt		Yard	Country	Delivery	Chartering
Hagb. Waage	472	TT	Howaldtswerke	Germany	1978	Unfixed
Hagb. Waage	320	TT	Howaldtswerke	Germany	1978	Unfixed
Hagb. Waage	258	TT	Aker Group	Norway	2 qtr 1974	Unfixed
Hagb. Waage	380	TT	Aker Group	Norway	December 1975	Unfixed
Hagb. Waage	380	TT	Aker Group	Norway	Nov 1977	Unfixed
H Ditlev-Simonsen	276	TT	Mitsui Shipbuilding	Japan	March 1978	Unfixed
H Ditlev-Simonsen	415	TT	Kockums Mekaniska	Sweden	December 1976	Unfixed
H Ditlev-Simonsen	410	TT	Kockums Mekaniska	Sweden	January 1978	Unfixed
Harry Borthen and Co	410	TT	Verolme	The Neth.	April 1975	Unfixed
HE Hansen-Tangen	267	TT	Uddevallavarvet	Sweden	June 1975	5 year/1980
HE Hansen-Tangen	267	TT	Uddevallavarvet	Sweden	1975/1976	Unfixed
Hilmar Reksten	410	TT	Aker Group	Norway	1 qtr 1974	Unfixed
Hilmar Reksten	414	TT	Aker Group	Norway	2 qtr 1975	Unfixed
Hilmar Reksten	280	TT	Aker Group	Norway	1975	Unfixed
Hilmar Reksten	380	TT	Aker Group	Norway	1975	Unfixed
Hilmar Reksten	360	TT	Aker Group	Norway	4 qtr 1977	Unfixed
Hilmar Reksten	380	TT	Aker Group	Norway	1978	Unfixed
Hilmar Reksten	285	TT	Aker Group	Norway	1978	Unfixed
Hilmar Reksten	285	TT	Aker Group	Norway	1978	Unfixed
Jørgen Bang	352	TT	Kockums Mekaniska	Sweden	May 1975	2 year/1976

Ship-owner	Dwt		Yard		Country	Delivery	Chartering
Knut Knutsen OAS	370	TT	Kawasaki Heavy I		Japan	December 1974	Unfixed
Knut Knutsen OAS	286	TT	Kawasaki Heavy I		Japan	December 1976	Unfixed
Mosvold Shipping AS	285	MT	Sumitomo		Japan	April 1975	Unfixed
Mosvold Shipping AS	370	TT	Sumitomo		Japan	End 1976	Unfixed
Odd Go-dager and Co.	285	TT	Uddevallavarvet		Sweden	June 1974	5 year/1979
Odd Go-dager and Co.	285	TT	Uddevallavarvet		Sweden	1st half 1976	Unfixed
Onstad Shipping AS	285	TT	Uddevallavarvet		Sweden	May 1977	Unfixed
P. Meyer	285	TT	Howaldtswerke		Germany	1977	Unfixed
P. Meyer	420	TT	Howaldtswerke (O/O)		Germany	3 qtr 1976	Unfixed
Peder Smedvig	420	TT	Howaldtswerke		Germany	1977	Unfixed
Sig. Bergesen dy and Co.	420	MT	Mitsui Shipbuilding		Japan	July 1974	6 year/1980
Sig. Bergesen dy and Co.	420	TT	Mitsui Shipbuilding		Japan	July 1975	7 year/1982
Sig. Bergesen dy and Co.	233	TT	Mitsui Shipbuilding		Japan	December 1975	7 year/1983
Sig. Bergesen dy and Co.	230	TT	Kawasaki Heavy I		Japan	September 1976	Unfixed

Ship-owner	Dwt		Yard	Country	Delivery	Chartering
Sig. Bergesen dy and Co.	355	MS	Uljanik (O/O)	Yugoslavia	May 1975	8 year/1982
Skibs-AS Belships Ltd.	355	TT	Bremer Vulkan	Germany	April 1975	Unfixed
Thor Dahl AS	355	TT	Mitsui Shipbuilding	Japan	July 1976	Unfixed
Thor Dahl AS	355	MT	Mitsui Shipbuilding	Japan	September 1976	Unfixed
Tschudi and Eitzen	230	TT	Uddevallavarvet	Sweden	1977	Unfixed
Wilh. Wilhelmsen	233	TT	Aker Group	Norway	February 1974	5 year/1979
Wilh. Wilhelmsen	230	TT	Nippon Kokan	Japan	December 1976	Unfixed
Wilh. Wilhelmsen	485	TT	Nippon Kokan	Japan	April 1975	Unfixed
Wilh. Wilhelmsen	228	TT	Nippon Kokan	Japan	March 1976	Unfixed

Notes: The Olsen list includes six vessels not registered in *Norwegian Shipping News*, three of which are Sanko vessels ordered by Mosvold. The other three were ordered by Onstad, Belstove and Höegh, but I have not been able to confirm from other sources whether these were actual contracts or options. The six vessels have therefore been omitted from the table.

Source: "Norwegian New Building Contracts as per 1st January 1974," *Norwegian Shipping News*, No. 2A (1974), 41-64; and Johan G. Olsen, "World Tanker Fleet of 200,000 Tons and over Including Combination Carriers and New-buildings."

Appendix 3
Tonnage as a Share of the Norwegian Fleet by Port of Registry, 1970-1987 (Grt)

	1970	1971	1972	1973	1974	1975	1976	1977	1978	1979	1980	1981	1982	1983	1984	1985	1986	1987	1987 int.
Arendal	3.2	3.1	3.4	3.2	4.2	4.3	4.4	3.4	3.3	3.6	4.1	4.1	4.4	4.7	4.9	4.8	1.9		1.7
Bergen	12.2	11.7	11.4	11.6	11.2	11.5	12.2	13.4	13.6	13.9	13.6	13.0	12.8	11.5	11.3	9.3	9.6	7.6	13.9
Bodø	0.1	0.1	0.1	0.1	0.1														
Drammen	0.4	0.6	0.1	0.2	0.2	0.2	0.2	0.2	0.2	0.3	0.2	0.2	0.2	0.0	0.0				0.1
Egersund	0.1	0.1	0.3	0.2	0.2	0.2	0.2	0.1											
Farsund	2.0	1.9	1.6	1.6	1.3	1.1	1.1	1.2	1.0	0.9	0.8	0.6	0.7	1.0	1.1	2.2	1.7	0.9	1.5
Flekkefjord	0.2	0.2	0.1	0.1											0.1				
Fredrikstad	0.2	0.2													0.1	0.1	0.2		0.1
Grimstad	2.7	2.7	2.6	2.3	2.4	2.5	2.2	2.3	2.4	1.8	1.8	1.9	2.1	2.5	2.1	2.3	2.3	2.9	6.2
Haugesund	3.4	3.7	3.2	3.6	3.6	3.5	3.6	3.5	3.4	3.6	3.5	2.9	2.8	2.7	3.4	1.0	1.1	1.6	1.1
Horten														0.2	0.2	0.2	0.1		
Kirkenes	0.1	0.1	0.0																
Kristiansand	8.0	7.8	7.7	8.2	8.0	8.1	9.0	9.0	9.3	9.1	8.8	8.9	7.6	8.5	8.7	9.8	12.1	11.0	9.5
Larvik	0.7	1.0	0.9	0.8	0.7	0.6	0.9	0.6	0.7	0.2	0.3	0.2	0.2	0.2	0.1	0.2	0.3	0.5	0.4
Lillesand	0.2	0.2	0.2	0.2	0.3	0.3	0.3	0.2	0.3	0.3	0.3	0.3	0.3	0.3	0.4	0.4			
Mandal	0.5	0.7	0.6	0.5	0.7	0.7	0.8	0.8	0.8	0.3									
Moss	1.5	1.4	1.3	1.5	1.2	0.8	1.8	1.8	1.7	1.7	1.9	1.9	0.1						
Narvik							0.0	0.0	0.0	0.0									
Oslo	37.8	36.3	38.3	39.5	38.7	38.9	36.4	36.8	36.6	37.5	37.7	38.5	38.9	37.9	34.8	36.2	31.6	34.0	59.7
Porsgrunn	0.7	0.6	0.6	0.6	0.4	0.4	0.3	0.3	0.4	0.5	0.5	0.5	0.5	0.5	0.7	0.8	0.6	1.0	1.1
Sandefjord	8.4	9.1	9.9	9.3	9.5	9.8	9.3	9.3	9.1	8.7	8.9	8.9	9.9	7.0	6.4	6.0	4.9	2.9	2.4
Skien											0.0	0.0	0.0	0.0	0.0				
Skudeneshavn	0.2	0.2	0.2	0.3	0.2	0.2	0.2	0.1	0.1	0.2	0.2	0.2	0.2						
Stavanger	9.4	9.5	9.2	9.0	10.1	10.4	10.3	10.5	9.9	11.4	11.7	12.3	14.5	17.4	19.6	21.3	25.4	29.0	1.9
Stokm.	0.1	0.1	0.1	0.1	0.1	0.1	0.1	0.1	0.1	0.1	0.1	0.1	0.1	0.1	0.1				
Tromsø	0.5	0.5	0.1	0.1	0.1	0.1	0.1	0.1	0.1								0.4		0.5

Stig Tenold

	1970	1971	1972	1973	1974	1975	1976	1977	1978	1979	1980	1981	1982	1983	1984	1985	1986	1987	1987 int.
Trondheim	0.4	0.2	0.2	0.2	0.3	0.2	0.2	0.2	0.3	0.2	0.3	0.3	0.1	0.2	0.2	0.2	0.4	0.5	0.1
Tvedestrand	0.1	0.1	0.1	0.0						0.0									
Tønsberg	6.5	7.8	7.0	6.4	6.4	6.3	6.7	6.2	6.8	5.7	5.4	5.3	4.7	5.3	5.8	5.0	7.5	7.5	0.2
Ålesund	0.5	0.4	0.5	0.5	0.3														

Notes: An entry of "0.0" implies that vessels above 5000 gross registered tons (grt) have been registered in the port, but that the tonnage represents less than .05 percent of the Norwegian fleet. The column "1987 int" refers to vessels registered abroad but managed by an owner in the port in question. The fleets of Sig. Bergesen d.y. and Wilh. Wilhelmsen have been assigned to Stavanger and Tønsberg, respecitvely.

Source: Database compiled on the basis of the Veritas register.

Select Bibliography

Abrahamsson, Bernhard J. "International Shipping: Developments, Prospects and Policy Issues." *Ocean Yearbook 8*. Chicago: University of Chicago Press, 1987, pp. 158-175.

_____. "Merchant Shipping in Transition." *Ocean Yearbook 4*. Chicago: University of Chicago Press, 1983, pp. 121-139.

Ahrari, Mohammed E. *OPEC: The Falling Giant*. Lexington: University Press of Kentucky, 1986.

Ait-Laoussine, Nordine and Parra, Francisco R. "The Development of Oil Supplies during the Energy Crises of the 1970s and Some Questions for the Future." *OPEC Review*, IX (1985).

Andersen, Håkon With. "Producing Producers: Shippers, Shipyards and the Cooperative Infrastructure of the Norwegian Maritime Complex since 1850." In Sabel, Charles F. and Zeitlin, Jonathan (eds.). *World of Possibilities: Flexibility and Mass Production in Western Industrialization*. Cambridge: Cambridge University Press, 1997, pp. 461-500.

_____ and Collett, John Peter. *Anchor and Balance. Det norske Veritas 1864-1989*. Oslo: J.W. Cappelens Forlag, 1989.

Andreassen, Jan A. Berg. "Risk and Investment Decisions in Non-liner Shipping." *Maritime Policy and Management*, XVII, No. 1 (1991), pp. 23-30.

_____. "Some Aspects of a Scrap-and-Build Scheme." *Norwegian Shipping News*, No. 21 (1978).

Aukrust, Odd. *Norges økonomi etter krigen* [*The Norwegian Economy after the War*]. Oslo: Central Bureau of Statistics, 1965.

Apostolides, Michael. "VLCC/ULCCs: The Dying Giants." *Maritime Policy and Management*, X, No. 2 (1983), pp. 87-91.

Bailey, Elizabeth E. and Baumol, William J. "Deregulation and the Theory of Contestable Markets." *Yale Journal on Regulation*, I, No. 2 (1984), pp. 111-137; reprinted in Burgess, Giles H., Jr. (ed.). *Antitrust and Regulation*. Aldershot: Edward Elgar, 1992, pp. 420-446.

Bakka, Dag Jr. *Byen ved de syv hav* [*The City by the Seven Seas*]. Bergen: Seagull Publishing, 1998.

_____. *Hav i storm og stille. AS HAV Helmer Staubo and Co 1915-1990* [*The History of the Shipowner Helmer Staubo's Company AS HAV*]. Larvik: Krohn Johansen Forlag, 1990.

_____. *Höegh. Shipping through Cycles*. Oslo: Leif Höegh and Co., 1997.

Bakke, Hallvard. "Norsk skipsfartspolitikk" ["Norwegian Shipping Policy"]. *Internasjonal Politikk*, No. 1B (1979), pp. 207-218.

Barry, David; Casson, Mark; and Horner, Dennis. "The Shipping Industry."
In Casson, Mark (ed.). *Multinationals and World Trade. Vertical Integration and the Diffusion of Labour in World Industries*. London:
Allen and Unwin, 1986, pp. 343-371.

Bauchet, Pierre. "Is the West Doomed to Lose Its Maritime-related Industries?" *Maritime Policy and Management*, XVII, No. 1 (1990), pp. 3-13.

Bayliss, B. "Raw Material Resources and Transport." Unpublished paper presented to the Sixth International Symposium on Theory and Practice in Transport Economics, Madrid, 1977.

Bedriftsøkonomen. Various issues, 1970-1985.

Beenstock, Michael. *A Theory of Ship Prices*. London: City University Business School, 1984.

_____ and Vergottis, Andreas. "An Econometric Model of the World Market for Dry Cargo Freight and Shipping." *Applied Economics*, XXI, No. 3 (1989), 339-356.

_____ and _____. "An Econometric Model of the World Tanker Market." *Journal of Transport Economics and Policy*, XXIII, No. 3 (1989), pp. 263-280.

_____ and _____. *Econometric Modelling of World Shipping*. London: Chapman and Hall, 1993.

Bennett, P.G. and Giesen, M.O. "Aristotle's Fallacy: A Hypergame in the Oil Shipping Business." *Omega*, VII, No. 4 (1979).

_____; Huxham, C.S.; and Dando, M.R. "Shipping in Crisis: A Trial Run for 'Live' Application of the Hypergame Approach." *Omega*, IX, No. 6 (1981), pp. 579-594.

Berggren, Lars. "The Effects of the Shipyard Crisis in Malmö, Southern Sweden." In Erstevåg, Randi; Starkey, David J.; and Austbø, Anne Tove (eds.). *Maritime Industries and Public Intervention*. Stavanger: Stavanger Maritime Museum, 1995, pp. 194-204.

Beth, Ludwig; Hader, Arnulf; and Kappel, Robert. *25 Years of World Shipping*. London: Fairplay Publications, 1984.

Bjerkholt, Olav; Offerdal, Erik; and Strøm, Steinar (eds.). *Olje og gass i norsk økonomi* [*Oil and Gas in the Norwegian Economy*]. Oslo: Universitetsforlaget, 1985.

Blanchard, Oliver Jean and Fischer, Stanley. *Lectures on Macroeconomics*. Cambridge, MA: MIT Press, 1989.

Blom, Per Gustav; Fjellstad, Knut Morten; and Nessen, Frode. "Investeringsatferd i norske rederier – en casestudie av rederiet Peder Smedvig A/S" ["Investment Behaviour in Norwegian Shipowning Companies – A Case Study of the Shipowning Company Peter Smedvig A/S"]. Unpublished thesis, Norwegian School of Economics and Business Administration, 1983.

Borgen, Erling. *Hilmar Rekstens eventyr* [*The Adventures of Hilmar Reksten*]. Oslo: J.W. Cappelens Forlag, 1981.

_____. *Huset Bergesen* [*The House of Bergesen*]. Oslo: J.W. Cappelens Forlag, 1984.

Borgen, Svein Ole and Spanne, Martin. "Investeringsatferd i norske rederier: en casestudie av rederiet Knut Knutsen O.A.S." ["Investment Behaviour in Norwegian Shipowning Companies – A Case Study of Knut Knutsen OAS"]. Unpublished siviløkonom thesis, Norwegian School of Economics and Business Administration, 1982.

Branch, Alan E. *Elements of Shipping*. London: Chapman and Hall, 1964; 6th ed., London: Chapman and Hall, 1989.

Brautaset, Camilla. "Norsk eksport 1830-1865 i perspektiv av historiske nasjonalregnskaper" ["Norwegian Exports 1830-1865 in an Historical National Accounts Perspective"]. Unpublished Dr. oecon. thesis, Norwegian School of Economics and Business Administration, 2002.

Breistein, Gisle Stray. *Valutaregulering og skipsfart* [*Currency Control and Shipping*]. Bergen: Institute for Shipping Research, Norwegian School of Economics and Business Administration, 1984.

British Petroleum. *BP Statistical Review of World Energy 1984*. London: British Petroleum, 1984.

Broeze, Frank. *The Globalisation of the Oceans: Containerisation from the 1950s to the Present*. St. John's: International Maritime Economic History Association, Research in Maritime History No. 22, 2002.

_____. "Containerization and the Globalization of Liner Shipping." In Starkey, David J. and Harlaftis, Gelina (eds.). *Global Markets: The Internationalization of the Sea Transport Industries since 1850*. St. John's: International Maritime Economic History Association, Research in Maritime History No. 14, 1998, pp. 385-423.

Bøe, Øystein and Hope, Einar. "Investment Behaviour in Norwegian Bulk Shipping." In Hope, Einar (ed.). *Studies in Shipping Economics in Honour of Professor Arnljot Strømme Svendsen*. Oslo: Bedriftsøkonomenes Forlag, 1981, pp. 43-54.

Böhme, Hans. *Restraint on Competition in World Shipping*. London: Trade Policy Research Centre, 1978.

Cafruny, Alan W. *Ruling The Waves: The Political Economy of International Shipping*. Berkeley: University of California Press, 1987.

_____. "The Political Economy of International Shipping: Europe versus America." *International Organization*, XXXIX (1985), pp. 79-119.

Cashman, John P. "A Worldwide View of Shipping and Shipbuilding." *Maritime Policy and Management*, XII, No. 2 (1985), pp. 113-134.

_____. "Shipping Statistics – Who Owns the Fleet?" In Strandenes, Siri Pettersen; Svendsen, Arnljot Strømme; and Wergeland, Tor (eds.). *Shipping Strategies and Bulk Shipping in the 1990s*. Bergen: Centre for International Business, Norwegian School of Economics and Business Administration, 1989, pp. 73-85.

Casson, Mark. "The Role of Vertical Integration in the Shipping Industry." *Journal of Transport Economics and Policy*, XX, No. 1 (1986), pp. 7-29.

_____ (ed.). *Multinationals and World Trade. Vertical Integration and the Diffusion of Labour in World Industries*. London: Allen and Unwin, 1986.

Charemza, W. and Gronicki, M. "An Econometric Model of World Shipping and Shipbuilding." *Maritime Policy and Management*, VIII, No. 1 (1981), pp. 21-30.

Chida, Tomohei. "The Development of Japan's Post-war Shipping Policy." *Journal of Transport History*, New ser., V, No. 1 (1984), pp. 82-90.

_____ and Davies, Peter N. *The Japanese Shipping and Shipbuilding Industries. A History of Their Modern Growth*. London: Athlone Press, 1990.

Cho, Dong Sung and Porter, Michael E. "Changing Global Industry Leadership: The Case of Shipbuilding." In Porter, Michael E. (ed.). *Competition in Global Industries*. Boston: Harvard Business School Press, 1986, pp. 541-567.

Chrzanowski, Ignacy. *An Introduction to Shipping Economics*. London: Fairplay Publications, 1985.

_____; Krzyzanowski, Maciel; and Luks, Krzysztof. *Shipping Economics and Policy. A Socialist View*. London: Fairplay Publications, 1979.

Corlett, Ewan. *The Ship. The Revolution in Merchant Shipping 1950-1980*. London: Her Majesty's Stationery Office, 1981.

Couper, A.D. *The Geography of Sea Transport*. London: Hutchinson, 1972.

_____. "Future International Maritime Transport Developments and the Law of the Sea." *Ocean Yearbook 6*. Chicago: University of Chicago Press, 1986, pp. 97-106.

Dabrowski, K. "Comments on the Mechanisms of the World Shipping Market." *Maritime Policy and Management*, VIII, No. 2 (1981), pp. 85-98.

Dahl, Ole Felix. "Med Bergesen som los og kaptein: økonomisk og teknologisk utvikling i norsk og internasjonal verftsindustri 1945-70 med spesiell vekt på A/S Rosenberg mekaniske verksted" ["Bergesen as Pilot and Captain – Economic and Technological Development in the Norwegian and International Heavy Industry 1945-70 with Special Emphasis on AS Rosenberg Mekaniske Verksted"]. Unpublished hovedoppgave thesis, University of Bergen, 1995.

De Voogd, Cees. "Public Intervention and the Decline of Shipbuilding in the Netherlands." In Erstevåg, Randi; Starkey, David J.; and Austbø, Anne Tove (eds.). *Maritime Industries and Public Intervention*. Stavanger: Stavanger Maritime Museum, 1995, pp. 240-254.

Devanney, J.W. "A Model of the Tanker Market and a Related Dynamic Program." In Lorange, Peter and Norman, Victor D. (eds.). *Shipping Management*. Bergen: AS John Grieg, 1973, pp. 100-113.

Dobler, Jean-Pierre. "The Use of Long-Term Forecasting in Maritime Economics." In Lorange, Peter, and Norman, Victor D. (eds.). *Shipping Management*. Bergen: AS John Grieg, 1973, pp. 57-75.

_____. "The World Shipbuilding Capacity." In Rydén, Inger and von Schirach-Szmigiel, Christopher (eds.). *Shipping and Ships for the 1990s*. Stockholm: Stockholm School of Economics, 1979, pp. 117-142.

Dobrowen, Kim, *et al. Rederi og kapitaltilførsel – Hovedproblemer i moderne rederifinansiering* [*Shipowners and Capital Supply – Main Problems in Modern Ship Financing*]. Oslo: Nordic Institute of Maritime Law, University of Oslo, 1981.

Douglas, Peter S. "Financial Policies for the Shipping Industry." In Rydén, Inger and von Schirach-Szmigiel, Christopher (eds.). *Shipping and Ships for the 1990s*. Stockholm: Stockholm School of Economics, 1979, pp. 166-178.

Dragesund, Petter. "Kontraheringsatferd i tankmarkedet" ["Contracting Behaviour in the Tanker Market"]. Unpublished høyere avdeling thesis, Norwegian School of Economics and Business Administration, 1990.

Drewry, H.P. Shipping Consultants Ltd. *Independent Tanker Owners*. London: H.P. Drewry Shipping Consultants Ltd., 1979.

_____. *Shipping Statistics and Economics*. London: H.P. Drewry Shipping Consultants Ltd., 1970.

_____. *Tanker Market Reports*. London: H.P. Drewry Shipping Consultants Ltd. Various years.

_____. *The Drewry Tanker Market Review*. London: H.P. Drewry Shipping Consultants Ltd., 1982.

_____. *The Role of Independent Tanker Owners*. London: H.P. Drewry Shipping Consultants Ltd., 1976.

_____. *The Tanker Crisis: Causes, Effects and Prospects to 1985*. London: H.P. Drewry Shipping Consultants Ltd., 1976.

_____. *The Tanker Fleets of the International Oil Companies*. London: H.P. Drewry Shipping Consultants Ltd., 1979.

_____. *The Trading Outlook for Very Large Tankers*. London: H.P. Drewry Shipping Consultants Ltd., 1975.

_____. *World Shipping under Flags of Convenience*. London: H.P. Drewry Shipping Consultants Ltd., 1975.

Drury, Charles and Stokes, Peter. *Ship Financing. The Credit Crisis: Can the Debt/Equity Balance Be Restored?* London: Lloyd's of London Press, 1983.

Du Jonchay, Yvan. *The Handbook of World Transport.* London: Macmillan, 1978.

Eckbo, Bjørn Espen. *Risikopreferansen blant noen skandinaviske tankredere før og etter krisen i tankmarkedet [Risk Preference among Some Scandinavian Tanker Owners before and after the Tanker Market Crisis].* Bergen: Institute for Shipping Research, Norwegian School of Economics and Business Administration, 1977.

Economist. Various issues.

Einarsen, Johan. *Reinvestment Cycles and Their Manifestation in the Norwegian Shipping Industry.* Oslo: Institute of Economics, University of Oslo, 1938.

Eriksen, Ib Erik. "The Demand for Bulk Ship Services." In Hope, Einar (ed.). *Studies in Shipping Economics in Honour of Professor Arnljot Strømme Svendsen.* Oslo: Bedriftsøkonomenes Forlag, 1981, pp. 55-62.

_____ and Norman, Victor D. *Ecotank - modell for analyse av tankmarkedenes virkemåte [Ecotank - Econometric Model for Tanker Companies].* Bergen: Institute for Shipping Research, Norwegian School of Economics and Business Administration, 1976.

_____ and _____. "Skipsfarten i norsk samfunnsøkonomi" ["Shipping in the Norwegian Economy"]. *Statsøkonomisk Tidsskrift,* LXXXVII, No. 3 (1973), pp. 129-131.

Espeli, Harald. *Industripolitikk på avveie - Motkonjunkturpolitikken og Norges Industriforbunds rolle 1975-80 [Industrial Policy Gone Astray - The Counter-cyclical Policy and the Role of Norges Industriforbund].* Oslo: Ad Notam Gyldendal, 1992.

Eyre, John L. "A Ship's Flag - Who Cares?" *Maritime Policy and Management,* XVI, No. 3 (1989), pp. 179-187.

Fairplay. Various years.

Farmand. Various issues, 1970-1980.

Farrell, Trevor M.A. "The World Oil Market, 1973-1983, and the Future of Oil Prices." *OPEC Review,* IX, No. 4 (1985), pp. 389-416.

Fearnley and Egers Chartering Co. Ltd. *Review.* Oslo: Fearnley and Egers Chartering Co. Ltd., 1966-1989.

Financial Times. *Financial Times World Shipping Yearbook, 1975-76.* London: Financial Times, 1976.

_____. Various issues.

Fischer, Lewis R. (ed.). *From Wheel House to Counting House; Essays in Maritime Business History in Honour of Professor Peter Neville Davies*. St. John's: International Maritime Economic History Association, Research in Maritime History No. 2, 1992.

_____ and Nordvik, Helge W. "Economic Theory, Information and Management in Shipborking: Fearnley and Eger as a Case Study, 1869-1972." In Ville, Simon P. and Williams, David M. (eds.). *Management, Finance and Industrial Relations in Maritime Industries: Essays in International Maritime and Business History*. St. John's: International Maritime Economic History Association, Research in Maritime History No. 6, 1994, pp. 1-29.

_____ and _____. "Subsidy and Protection in National Shipping Industries around the North Sea since World War II." In Erstevåg, Randi; Starkey, David J.; and Austbø, Anne Tove (eds.). *Maritime Industries and Public Intervention*. Stavanger: Stavanger Maritime Museum, 1995, pp. 84-101.

_____ and Panting, Gerald D. (eds.). *Change and Adaptation in Maritime History: The North Atlantic Fleets in the Nineteenth Century*. St. John's: Maritime History Group, Memorial University of Newfoundland, 1985.

Fon, Anders Martin. "En stormakt i tørrbulk. En økonomisk-historisk analyse av norsk tørrbulkfart 1950-1973" ["A Major Dry Bulk Nation: An Economic Historical Analysis of Norwegian Dry Bulk Shipping 1950-1973"]. Unpublished Dr. oecon. thesis, Norwegian School of Economics and Business Administration, 1995.

_____. "Norsk og gresk skipsfartsnæring 1945-1995 – en sammenligning" ["Norwegian and Greek Shipping 1945-1995 – A Comparison"]. *Sjøfartshistorisk Årbok 1995* [*Norwegian Yearbook of Maritime History 1995*]. Bergen: Bergen Maritime Museum, 1996, pp. 49-75.

_____. "Two Markets or One? An Historical Study of Price Behaviour in the Tanker and Dry Bulk Shipping Markets, 1955-1973." *International Journal of Maritime History*, VII, No. 2 (1995), pp. 115-134.

Forsyth, Craig J. "Transnational Corporations: Problems for Study in the New International Order of Maritime Shipping." *Maritime Policy and Management*, XX, No. 3 (1993), pp. 207-214.

Franck-Nielsen, Arne. *Redere – tank – olje* [*Shipowners – Tankers – Oil*]. Bergen: Study of the Distribution of Power in Norway, 1975.

Frankel, Ernst G. *The World Shipping Industry*. London: Croom Helm, 1987.

_____. "Shipping and Its Role in Economic Development." *Marine Policy*, XIII, No. 1 (1989), pp. 22-42.

_____, *et al. Bulk Shipping and Terminal Logistics*. Washington, DC: World Bank, 1989.

Gardner, B.M.; Pettit, S.J.; and Thanopoulou, H.A. "Shifting Challenges for British Maritime Policy: A Post-war Review." *Marine Policy*, XX, No. 6 (1996), pp. 517-524.

General Agreement on Tariffs and Trade. *International Trade*. Geneva: General Agreement on Tariffs and Trade. Various years.

Glen, David. "The Emergence of Differentiation in the Oil Tanker Market, 1970-1978." *Maritime Policy and Management*, XVII, No. 4 (1990), pp. 289-312.

_____; Owen, M.; and van der Meer, R. "Spot and Time Charter Rates for Tankers 1970-77." *Journal of Transport Economics and Policy*, XV, No. 1 (1981), pp. 45-58.

Gombrii, Karl Johan, *et al. Nordisk skipsfart under fremmed flagg* [*Nordic Shipping Flying Foreign Flags*]. Oslo: Nordic Institute of Maritime Law, University of Oslo, 1981.

Gordon, Richard L.; Jacoby, Henry D.; and Zimmerman, Martin B. (eds.). *Energy: Markets and Regulation: Essays in Honor of M.A. Adelman*. Cambridge, MA: MIT Press, 1987.

Goss, Richard O. *Studies in Maritime Economics*. Cambridge: Cambridge University Press, 1968.

_____. "A Comment on Risk Preference and Shipping Decisions." In Lorange, Peter and Norman, Victor D. (eds.). *Shipping Management*. Bergen: AS John Grieg, 1973, pp. 182-188.

_____. "Economics and the International Regime for Shipping." *Maritime Policy and Management*, XI, No. 2 (1984), pp. 135-145.

_____. "Rochdale Remembered." *Maritime Policy and Management*, XXV, No. 3 (1998), pp. 213-233.

Goto, Shin. "Competitive Advantage in the Japanese Shipbuilding Industry: The Case of IHI." In Yuzawa, Takeshi (ed.). *Japanese Business Success. The Evolution of a Strategy*. London: Routledge, 1994.

_____. "Globalization and International Competitiveness: The Experience of the Japanese Shipping Industry since the 1960s." In Starkey, David J. and Harlaftis, Gelina (eds.). *Global Markets: The Internationalization of the Sea Transport Industries since 1850*. St. John's: International Maritime Economic History Association, Research in Maritime History No. 14, 1998, pp. 355-384.

Gram, Per. "Kansellering av certepartier" ["Cancellation of Charter Parties"]. *Norwegian Shipping News*, Nos. 12/13 (1975).

Groth, Vibecke and Lund, Jørgen. *Rederifinansiering og Statsgaranti* [*Ship Financing and Government Guarantees*]. Oslo: Nordic Institute of Maritime Law, University of Oslo, 1978.

Hambros Bank. *Report of the Directors and Accounts*. Various years.

Hammer, Jan A. and Høy, Arvid. *Norsk skipsfart og dets lånekilder* [*Norwegian Shipping and Its Sources of Finance*]. Bergen: Institute for Shipping Research, Norwegian School of Economics and Business Administration, 1983.

Hanisch, Tore Jørgen and Bould, Martin. *Megler og Reder – Torvald Klaveness gruppen 1946-1996* [*Shipbroker and Shipowner – The Torvald Klaveness Group 1946-1996*]. Oslo: Messel Forlag, 1996.

_____ and Nerheim, Gunnar. "Fra vantro til overmot" ["From Disbelief to Arrogance"]. In *Norwegian Oil History – Volume 1*. Oslo: Norsk Petroleumsforening, 1992.

_____ and Ramskjær, Liv Jorunn. *Firmaet Sigval Bergesen, Stavanger: under vekslende vilkår 1887-1987*. Stavanger: Dreyer bok, 1987.

Hansen, Arild Marøy and Thowsen, Atle. *Sjøfartshistorie som etterkrigshistorisk forskningsfelt.* [*Postwar Maritime History as a Topic for Research*]. Bergen: Norwegian Research Centre in Organization and Management, University of Bergen, 1991.

Hansen, Terje. *Aarbakke-utvalgets skisse til reform av bedrifts og kapitalbeskatningen og norsk skipsfartsnæring* [*The Aarbakke Committee's Sketch for Reform of Company and Capital Taxation and the Norwegian Shipping Industry*]. Bergen: Centre for Applied Research, Norwegian School of Economics and Business Administration, 1989.

Harlaftis, Gelina, *Greek Shipowners and Greece 1945-1975*. London: Athlone Press, 1993.

_____. *A History of Greek-owned Shipping: The Making of an International Tramp Fleet, 1830 to the Present Day*. London: Routledge, 1996.

Hauge, Odd Harald. *Fred. Olsen – uautorisert biografi* [*Fred. Olsen – An Unauthorised Biography*]. Oslo: Gyldendal Norsk Forlag, 1993.

_____ and Stavrum, Gunnar. *John Fredriksen – uautorisert biografi* [*John Fredriksen – An Unauthorised Biography*]. Oslo: Gyldendal Norsk Forlag, 1991.

Hawdon, D. "Tanker Freight Rates in the Short and Long Run." *Applied Economics*, X, No. 3 (1978), 203-217.

Hirsch, H.J. Darre. "Rederorganisasjonene og internasjonalt samarbeid." In Sjaastad, Anders C. (ed.). *Norsk Utenrikspolitisk Årbok 1975*. Oslo: Norwegian Institute of International Affairs, 1976, pp. 99-112.

Hodne, Fritz. *Norges økonomiske historie 1815-1970* [*The Economic History of Norway, 1815-1970*]. Oslo: J.W. Cappelens Forlag, 1981.

Hole, Vidar. "The Biggest Gamblers: Structure and Strategy in Norwegian Oil Tanker Shipping during the Era of Growth, 1925-1973." Unpublished høyere avdeling thesis, Norwegian School of Economics and Business Administration, 1993.

Hope, Einar (ed.). *Studies in Shipping Economics in Honour of Professor Arnljot Strømme Svendsen*. Oslo: Bedriftsøkonomenes Forlag, 1981.

Hope, Ronald. *A New History of British Shipping*. London: John Murray, 1990.

Horringmo, H. "Some Aspects of Ship Financing." *Norwegian Shipping News*, No. 8B (1973).

Jenkins, Gilbert; Stopford, Martin; and Tyler, Cliff. *The Clarkson Oil Tanker Databook*. London: Clarkson, 1993.

Johnman, Lewis. "Internationalization and the Collapse of British Shipbuilding, 1945-1973." In Starkey, David J. and Harlaftis, Gelina (eds.). *Global Markets: The Internationalization of the Sea Transport Industries since 1850*. St. John's: International Maritime Economic History Association, Research in Maritime History No. 14, 1998, pp. 319-354.

_____. "Public Intervention and the 'Hollowing Out' of British Shipbuilding: The Swan Hunter Closure." In Erstevåg, Randi; Starkey, David J.; and Austbø, Anne Tove (eds.). *Maritime Industries and Public Intervention*. Stavanger: Stavanger Maritime Museum, 1995, pp. 223-239.

Jonasen, Jonas Schanche. *Sigval Bergesen 1863-1956*. Stavanger: Sigval Bergesen, 1963.

Joplin, A.F. "Development of Product Tankers in World Trade." In Second International Waterborne Transportation Conference *Proceedings*. New York: American Society of Civil Engineers, 1977.

Kapital. Various years, 1970-1985.

Karlsen, Jan Karl. *The Mobile Rig Industry*. Bergen: Foundation for Research in Economics and Business Administration, Norwegian School of Economics and Business Administration, 1992.

Kaukiainen, Yrjö. *A History of Finnish Shipping*. London: Routledge, 1993.

Kendall, Lane C. *The Business of Shipping*. Cambridge, MD: Cornell Maritime Press, 1973; 5th ed., Centreville, MD: Cornell Maritime Press, 1986.

Knudsen, Kristen. "The Economics of Zero Taxation of the World Shipping Industry." *Maritime Policy and Management*, XXIV, No. 1 (1997), pp. 45-54.

Kolltveit, Bård. *The Golar Chronicle: Gotaas-Larsen Shipping Corporation 1946-1991*. Oslo: Gotaas-Larsen, 1991.

_____. *Bridge across the Seas: AS Ivarans Rederi 1920-1995*. Oslo: AS Ivarans Rederi, 1995.

_____; Vadseth, Knut; and Butenschøn, Hans B. *Six Decades on the Seven Seas: A Saga of Value Creation. The Anders Wilhelmsen Group 1939-1999*. Oslo: Andresen and Butenschøn, 2000.

Koopmans, Tjalling C. *Tanker Freight Rates and Tankship Building: An Analysis of Cyclical Fluctuations*. London: P. S. King and Son, 1939.

Kuuse, Jan. "The Relations between the Swedish Shipbuilding Industry and Other Industries 1900-1980." In Kuuse, Jan and Slaven, Anthony (eds.). *Development Problems in Historical Perspective*. Glasgow: Scottish and Scandinavian Shipbuilding Seminar, 1980.

_____ and _____(eds.). *Development Prospects in Historical Perspective.* Glasgow: Scottish and Scandinavian Shipbuilding Seminar, 1980.

Lensberg, Terje and Rasmussen, Heine. *A Stochastic Intertemporal Model of the Tanker Market.* Bergen: Foundation for Research in Economics and Business Administration, Norwegian School of Economics and Business Administration, 1992.

Lorange, Peter and Norman, Victor D. (eds.) *Shipping Management.* Bergen: AS John Grieg, 1973.

_____ and _____. "How Attitudes towards Risk Influence Investment Decisions." *European Business,* No. 33 (1972), pp. 71-84.

_____ and _____. "Risk Preference in Scandinavian Shipping." *Applied Economics,* V, No. 1 (1973), pp. 49-59.

Lorentzen, Irene Cecilie. "The Internationalisation of Norwegian Shipping Companies." Unpublished mss., Kristiansand, 1995.

Lund, Ole. "Shippingkrise og stat" ["Shipping Crisis and the State"]. *Bergen Bank Kvartalsskrift* [*Bergen Bank Quarterly Journal*], No. 2 (1985), pp. 53-68.

_____. "Skipsfartskrise og konkurslov" ["Shipping Crisis and Bankruptcy Law"]. *Bergen Bank Kvartalsskrift* [*Bergen Bank Quarterly Journal*], No. 3 (1988), pp. 57-64.

Løddesøl, Leif Terje. "Hvorfor gjør noen rederier det godt og andre det dårlig?" ["Why Are Some Shipping Companies Successful, and Some Not?"]. *Internasjonal Politikk,* No. 1B (1979), pp. 167-174.

Metaxas, Basil N. *The Economics of Tramp Shipping.* London: Athlone Press, 1971.

_____. *Flags of Convenience, A Study of Internationalisation.* London: Gower, 1985.

_____. "Notes on the Internationalisation Process in the Maritime Sector." *Maritime Policy and Management,* V, No. 1 (1978), pp. 51-61.

Mjelva, Hans K. "Stord Verft 1945-1975." *Sjøfartshistorisk Årbok 1995* [*Norwegian Yearbook of Maritime History 1995*]. Bergen: Bergen Maritime Museum, 1996, pp. 77-275.

_____. "Tre storverft i norsk industris finaste stund: ein komparativ studie av Stord Verft, Rosenberg mek. Verksted og Fredrikstad mek. Verksted 1960-1980" ["Three Major Yards in the Finest Hour of Norwegian Manufacturing: A Comparative Analysis of Stord Verft, Rosenberg mek. Verksted and Fredrikstad mek. Verksted 1960-1980"]. Unpublished PhD thesis, University of Bergen, 2005.

Mottershead, Peter. "Shipbuilding: Adjustment-Led Intervention or Intervention-Led Adjustment." In Shepherd, Geoffrey; Duchene, François; and Saunders, Christopher (eds.). *Europe's Industries: Public and Private Strategies for Change.* London: Francis Pinter, 1983.

Mäkinen, Eva Helena. *Shipping Economics: A Strategic Analysis.* Turku, 1986.

Nerheim, Gunnar; Jøssang, Lars Gaute; Utne, Bjørn S.; and Dahlberg, Frida. *I vekst og forandring – Rosenberg Verft 100 år 1896-1996* [*Growth and Change – Rosenberg Verft 100 years 1896-1996*]. Stavanger: Kværner Rosenberg, 1995.

_____ and Utne, Bjørn S. *Under samme stjerne – Rederiet Peder Smedvig 1915-1990* [*Under the Star – The Shipowning Company Peder Smedvig 1915-1990*]. Stavanger: Peder Smedvig A/S, 1990.

_____ and Gjerde, Kristin Øye. *Uglandrederiene – verdensvirksomhet med lokale røtter* [*The Ugland Shipowning Companies – International Business with Local Roots*]. Grimstad: Andreas K.L. Ugland and Johan Jørgen Ugland, 1996.

Nersesian, Roy. *Ships and Shipping – A Comprehensive Guide*. Tulsa, OK: PennWell Books, 1981.

Nordisk Skibsrederforening Medlemsblad. Various years, 1970-1980.

Nordvik, Helge W. "From Benign Neglect to Active Intervention: Norwegian Government Shipping Policies from the 1970s Shipping Crisis to the Present." Unpublished paper presented to the IAME Conference, City University, London, September 1997.

_____. "Norwegian Maritime Historical Research during the Past Twenty Years: A Critical Survey." *Sjøfartshistorisk Årbok 1990* [*Norwegian Yearbook of Maritime History 1990*]. Bergen: Bergen Maritime Museum, 1991, pp. 241-278.

_____. "The Norwegian Shipbuilding Industry – The Transition from Wood to Steel 1880-1980." In Walker, Fred M. (ed.). *European Shipbuilding: One Hundred Years of Change*. London: Marine Publications International, 1983.

_____. "The Shipping Industries of the Scandinavian Countries, 1850-1913." In Fischer, Lewis R. and Panting, Gerald D. (eds.). *Change and Adaptation in Maritime History: The North Atlantic Fleets in the Nineteenth Century*. St. John's: Maritime History Group, Memorial University of Newfoundland, 1985, pp. 117-148.

Norges Handels og Sjøfarts Tidende. Various issues, 1970-1980.

Norges Industri. Various years.

Norges Offentlige Utredninger (1983:13). *Reksten-saken* [*Report to the Norwegian Government from the Investigative Committee Established by Royal Resolution June 26, 1981 to go through the Reksten Case*].

_____ (1983:7). *Skipsfartens konkurranseevne* [*The Competitiveness of Shipping*].

_____ (1978:13). *Skipsfartsnæringen* [*The Shipping Sector*].

_____ (1980:45). *Uteregistrering av skip og skipsfartens egenkapital* [*Foreign Registry of Ships and the Equity of Shipping*].

Norges Rederforbund. *Momenter til belysning av norsk og internasjonal skipsfart* [*Aspects Illuminating Norwegian and International Shipping*]. Oslo. Various years.

_____. *Årsberetning* [*Annual Report*]. Various years.

Norman, Victor D. *Internasjonal Sjøtransport* [*International Sea Transport*]. Bergen: Institute for Shipping Research, Norwegian School of Economics and Business Administration, 1973.

_____. "Har det vært for lett å drive internasjonal skipsfart i Norge?" ["Has Shipping Been Too Easy an Enterprise in Norway?"], *Internasjonal Politikk*, No. 1B (1979), pp. 175-186.

_____. "Market Strategies in Bulk Shipping." In Hope, Einar (ed.). *Studies in Shipping Economics in Honour of Professor Arnljot Strømme Svendsen*. Oslo: Bedriftsøkonomens Forlag, 1981, pp. 13-28.

_____. *Norwegian Shipping in the National Economy*. Bergen: Institute for Shipping Research, Norwegian School of Economics and Business Administration, 1971.

_____. "Shipping Problems – Has the Market Mechanism Failed?" *Norwegian Shipping News*, No. 7 (1976).

_____. *The Economics of Bulk Shipping*. Bergen: Institute for Shipping Research, Norwegian School of Economics and Business Administration, 1979.

_____. "Skipsfartskrisen etter 1973 – forløp og årsaker" ["The Shipping Crisis after 1973 – Course and Causes"]. *Sjøfartshistorisk Årbok 1998* [*Norwegian Yearbook of Maritime History 1998*]. Bergen: Bergen Maritime Museum, 1999, pp. 161-182.

_____. *A Portfolio Selection Model of Shipping Behaviour*. Bergen: Institute for Shipping Research, Norwegian School of Economics and Business Administration, 1971.

Norway. Parliament. Innst. S. No. 167 (1978-1979). "Innstilling fra den forsterkede finanskomite om støttetiltak for skipsfartsnæringen" ["Recommendation from the Extended Committee on Finance and Economic Affairs on Support Measures for the Shipping Industry"].

_____. _____. Odeltingsproposisjon No. 45 (1986-1987). "Om lov om norsk internasjonalt skipsregister" ["On the Law on a Norwegian International Registry for Ships"].

_____. _____. "Rapport til Stortinget fra den granskningskommisjon i Reksten-saken som ble oppnevnt ved Stortingets vedtak 20. juni 1985" ["Report to the Storting from the Investigative Committee in the Reksten Case Established by Royal Resolution 20 June 1985"], 9 February 1988.

_____. _____. Stortingsmelding No. 8 (1979-1980). "Om virksomheten i Norsk Garantiinstitutt for skip og borefartøyer A/S i 1978" ["On the Activities of the Norwegian Guarantee Institute for Ships and Drilling Vessels Ltd. in 1978"].

_____. _____. Stortingsmelding No. 23 (1975-1976). "Om sjøfolkenes forhold og skipsfartens plass i samfunnet" ["On the Conditions for Seamen and Shipping's Role in the Society"].

_____. _____. Stortingsmelding No. 25 (1973-1974). "Petroleumsvirksomhetens plass i det norske samfunn" ["The Role of the Petroleum Business in Norwegian Society"].

_____. _____. Stortingsmelding No. 52 (1980-1981). "Om skipsfartsnæringen" ["On the Shipping Industry"].

_____. _____. Stortingsmelding No. 53 (1984-1985). "Om skipsfartsnæringen" ["On the Shipping Industry"].

_____. _____. Stortingsmelding No. 53 (1981-1982). "Om avtale av 5. februar 1982 mellom Norsk garantiinstitutt for skip og borefartøyer AS, Hambros Bank Limited, R/A Hadrian og R/A Trajan i forbindelse med avviklingen av garantiengasjementene i Rekstenflåten" ["On the Agreement dated 5 February 1982 between the Norwegian Guarantee Institute for Ships and Drilling Vessels Ltd., Hambros Bank Ltd., R/A Hadrian and R/A Trajan on the Winding-up of the Guarantee Engagements in the Reksten Fleet"].

_____. _____. Stortingsmelding No. 71 (1972-1973). "Langtidsprogrammet 1974-1977" ["Long-term Programme, 1974-1977"].

_____. _____. Stortingsproposisjon No. 1 (2005-2006). "Statsbudsjettet – Nærings- og Handels-departementet" ["The National Budget – Proposition from the Ministry of Trade and Industry"].

_____. _____. Stortingsproposisjon No. 17 (1975-1976) "Om etablering av en midlertidig garantiordning for norske skip og borerigger" ["Royal Proposition on the Establishment of a Temporary Guarantee Scheme for Norwegian Ships and Drilling Vessels"].

_____. _____. Stortingsproposisjon No. 21 (1982-1983). "Om bevilgning til utbetaling under garantier stillet av Norsk Garantiinstitutt for skip og borefarttøyer AS for lån til rederiet Knut Knutsen OAS" ["On the Appropriation of Pay-outs with Regard to Guarantees Granted by the Norwegian Guarantee Institute for Ships and Drilling Vessels Ltd. in Connection with Loans to the Shipowning Company Knut Knutsen OAS"].

_____. _____. Stortingsproposisjon No. 38 (1981-1982). "Om bevilgning til utbetaling under garanti stillet av Norsk Garantiinstitutt for skip og borefarttøyer" ["On the Appropriation of Pay-out with Regard to Guarantees Granted by the Norwegian Guarantee Institute for Ships and Drilling Vessels Ltd."].

_____. _____. Stortingsproposisjon No. 46 (1978-1979). "Om støttetiltak for skipsfartsnæringen" ["On Support Measures for the Shipping Industry"].

_____. _____. Stortingsproposisjon No. 101 (1976-1977). "Om tiltak på skipsbyggingssektoren" ["On Measures for the Shipbuilding Sector"].

_____. _____. Stortingsproposisjon No. 186 (1975-1976). "Om utviding av statens garantiansvar overfor Norsk garantiinstitutt for skip og bore-fartøyer A/S" ["Royal Proposition on the Expansion of the Authorities' Guarantee Responsibility towards the Norwegian Guarantee Institute for Ships and Drilling Vessels Ltd."].

_____. _____. Stortingsproposisjon No. 187 (1974-1975). "Om fullmakt til kjøp av aksjer i norske selskaper" ["On the Authorisation of the Purchase of Shares in Norwegian Companies"].

_____. Statistisk Sentralbyrå, *Historisk Statistikk 1994* [*Historical Statistics 1994*]. Oslo: Statistisk Sentralbyrå, 1994.

_____. _____. *Sjøtransport* [*Maritime Statistics*]. Oslo: Statistisk Sentralbyrå. Various years.

_____. _____. *Statistisk Månedshefte* [*Monthly Bulletin of Statistics*]. Oslo: Statistisk Sentralbyrå. Various years.

Norwegian Shipping News. Various issues, 1970-1980.

Næss, Erling Dekke. *Autobiography of a Shipping Man*. London: Seatrade, 1977.

_____. "Tankfartens problemer og utsikter" ["The Problems and Prospects of Tanker Shipping"]. Kristofer Lehmkuhl Lecture, Norwegian School of Economics and Business Administration, 1965.

Organisation for Economic Co-operation and Development. *Profits and Rates of Return*. Paris: Organisation for Economic Co-operation and Development, 1979.

_____. *Structural Adjustment and Economic Performance*. Paris: Organisation for Economic Co-operation and Development, 1987.

_____. *The Industrial Policies of 14 Member Countries*. Paris: Organisation for Economic Co-operation and Development, 1971.

_____. *Towards Full Employment and Price Stability*. Paris: Organisation for Economic Co-operation and Development, 1977.

_____. *OECD Economic Outlook: Historical Statistics 1960-1983*. Paris: Organisation for Economic Co-operation and Development, 1985.

_____. *OECD Economic Outlook: Historical Statistics 1960-1990*. Paris: Organisation for Economic Co-operation and Development, 1992.

_____. *Maritime Transport*, 1970-1987.

Onarheim, Onar. *Min tørn* [*My Bout*]. Oslo: Gyldendal Norsk Forlag, 1984.

Organisation of Petroleum Exporting Countires. *OPEC 1989 Statistical Bulletin*. Vienna: Organisation of Petroleum Exporting Countries, 1989.

Peters, Hans J. *Seatrade, Logistics and Transport*. Washington, DC: World Bank, 1989.

_____. *The Maritime Transport Crisis*. Washington, DC: World Bank, 1993.

Platou. *The Platou Report*. Various issues.

Porter, Michael E. "The Oil Tanker Shipping Industry." In Porter, Michael E. (ed.). *Cases in Competitive Strategy*. New York: Free Press, 1983, pp. 49-70.

_____. (ed.). *Cases in Competitive Strategy*. New York: Free Press, 1983.

_____. (ed.). *Competition in Global Industries*. Boston: Harvard Business School Press, 1988.

Rafgård, Tormod, "Oil Transportation in Tankers – Getting Cheaper and Cheaper." *Norwegian Shipping News*, No. 21 (1979).

Ratcliffe, Mike. *Liquid Gold Ships. A History of the Tanker 1859-1984*. London: Lloyd's of London Press, 1985.

Rederiaksjeselskapet Hadrian. *Driftsberetning og Resultat [Annual Reports]*. Various years.

Reksten, Hilmar. *Opplevelser [Experiences]*. Oslo: H. Aschehoug og Co., 1979.

_____. "Sig. Bergesen d.y. and Co. 1947-1970." Manuscript sent to Sig. Bergesen d.y. and Norwegian newspapers, 1971.

_____. "Noen ideer om konkurransevilje og risikomomentet under strukturendringene i norsk tankskipsfart" ["Some Ideas about Willingness to Compete and the Risk Aspect during the Structural Changes in Norwegian Tanker Shipping"]. Kristofer Lehmkuhl Lecture, Norwegian School of Economics and Business Administration, 1971.

Rinman, Thorsten and Broderfors, Rigmor. *The Commercial History of Shipping*. Gothenburg: Rinman and Lindén, 1983.

Røpke, Inge. "Den internationale krise i skibsfart og værftsindustri i 1970-erne" ["The International Crisis in Shipping and the Shipbuilding Industry in the 1970s"]. Unpublished spesialoppgave thesis, 2 vols., University of Copenhagen, 1981.

Seland, Johan. *Norsk skipsfart år for år 1946-1976 [Norwegian Shipping Year by Year, 1946-1976]*. Bergen: Fagbokforlaget, 1994.

Shepherd, Geoffrey; Duchene, François; and Saunders, Christopher (eds.). *Europe's Industries. Public and Private Strategies for Change*. London: Francis Pinter, 1983.

Shimojo, Tetsuji. *Economic Analysis of Shipping Freights*. Kobe: Research Institute for Economics and Business Administration, 1979.

Sig. Bergesen d.y. and Co. *Sigval Bergesen d.y. 27. april 1893-7. mai 1980*. Oslo: Sig. Bergesen d.y. and Co., 1980.

_____. *Report and Account*. Various years.

Sintef. *Norsk offshoreindustris konkurranseevne* [*The Competitiveness of the Norwegian Offshore Industry*]. Oslo: report delivered to the Ministry of Petroleum and Energy, 1985.

Slaven, Anthony. "Management Policy and the Eclipse of British Shipbuilding." In Walker, Fred M. (ed.). *European Shipbuilding: One Hundred Years of Change*. London: Marine Publications International, 1983.

_____. "Marketing Opportunities and Marketing Practices: The Eclipse of British Shipbuilding, 1957-1976." In Fischer, Lewis R. (ed.). *From Wheel House to Counting House: Essays in Maritime Business History in Honour of Professor Peter Neville Davies*. St. John's: International Maritime Economic History Association, Research in Maritime History No. 2, 1992, pp. 125-152.

Sletmo, Gunnar K. "Shipping's Fourth Wave: Ship Management and Vernon's Trade Cycles." *Maritime Policy and Management*, XVI, No. 4 (1989), pp. 293-303.

_____. and Holste, Susanne. "Shipping and the Competitive Advantage of Nations: The Role of International Ship Registers." *Maritime Policy and Management*, XX, No. 3 (1993), pp. 243-255.

Sohmen, Helmut, "Profitability in Shipping." *Kieler Vorträge gehalten im Institut für Weltwirschaft an der Universität Kiel*, CIII (1983).

Sosialøkonomen. Various issues, 1970-1985.

Spero, Joan Edelman. *The Politics of International Economic Relations*. London: St. Martin's Press, 1990.

Spruyt, John. *Ship Management*. London: Lloyd's of London Press, 1994.

Starkey, David J. and Harlaftis, Gelina (eds.). *Global Markets: The Internationalization of the Sea Transport Industries since 1850*. St. John's: International Maritime Economic History Association, Research in Maritime History No. 14, 1998.

Stokes, Peter. *Ship Finance: Credit Expansion and the Boom-Bust Cycle*. London: Lloyd's of London Press, 1992.

Stopford, Martin. *Maritime Economics*. London: Unwin Hyman, 1988; 2nd ed., London: Routledge, 1988.

_____. "Challenges and Pitfalls of Maritime Forecasting in a Corporate Environment." In Strandenes, Siri Pettersen; Svendsen, Arnljot Strømme; and Wergeland, Tor (eds.). *Shipping Strategies and Bulk Shipping in the 1990s*. Bergen: Centre for International Business, Norwegian School of Economics and Business Administration, 1989, pp. 39-48.

Strandenes, Siri Pettersen. *Kontrahering og salg av norske tankskip 1963-76* [*Contracting and Sales of Norwegian Tankers, 1963-76*]. Bergen: Institute for Shipping Research, Norwegian School of Economics and Business Administration, 1979.

_____. "Trekk ved konjunkturutviklingen i tankfarten" ["Aspects of Business Cycle Development in Tanker Shipping"]. Unpublished høyere avdeling thesis, Norwegian School of Economics and Business Administration, 1977.

_____; Svendsen, Arnljot Strømme; and Wergeland, Tor (eds.). *Shipping Strategies and Bulk Shipping in the 1990s*. Bergen: Centre for International Business, Norwegian School of Economics and Business Administration, 1989.

Stråth, Bo, "Industrial Restructuring in the Swedish Shipbuilding Industry." *Labour and Society*, XIV, No. 2 (1989), pp. 105-120.

Sturmey, S.G. *British Shipping and World Competition*. London: Athlone Press, 1962.

Svendsen, Arnljot Strømme. "Er rederbegavelsenes tid forbi?" ["Is the Era of Shipowner Talents Over?], *Bedriftsøkonomen*, No. 7 (1981), pp. 352-368.

_____. *Prospective Market Trends in the Shipping Trade*. Bergen: Institute for Shipping Research, Norwegian School of Economics and Business Administration, 1973.

_____. "Skipsfartskonjunkturene i 1970-årene" ["Shipping Cycles in the 1970s"]. *Sjøfartshistorisk Årbok 1978* [*Norwegian Yearbook of Maritime History 1978*]. Bergen: Bergen Maritime Museum, 1979, pp. 205-242.

_____. *Today's Great Paradox: The Wealth of the Sea and the Shipping Crises*. Bergen: Institute for Shipping Research, Norwegian School of Economics and Business Administration, 1977.

Sweden. Department of Communication. Statens offentliga utredningar (1976:44). *Sjøfart och flagg*,

Taylor, A.J. "Chartering Strategies for Shipping Companies." *Omega*, X, No. 1 (1982), pp. 25-33.

Tenold, Stig. *Skipsfartskrisen og norske redere – en økonomisk-historisk studie 1973-1980* [*The Shipping Crisis and Norwegian Shipowners – An Economic-Historical Study 1973-1980*]. Bergen: Foundation for Research in Economics and Business Administration, Norwegian School of Economics and Business Administration, 1995.

_____. *Norwegian Shipowning Companies and Foreign Direct Investment*. Bergen: Foundation for Research in Economics and Business Administration, Norwegian School of Economics and Business Administration, 2000.

_____. "The Shipping Crisis of the 1970s: Causes, Effects and Implications for Norwegian Shipping." Unpublished Dr. Oecon. thesis, Norwegian School of Economics and Business Administration, 2001.

_____. "Saving a Sector – But Which One? The Norwegian Guarantee Institute for Ships and Drilling Vessels Ltd." *International Journal of Maritime History*, XIII, No. 1 (June 2001), pp. 39-62.

_____. "The Harder They Come...Hilmar Reksten from Boom to Bankruptcy." *The Northern Mariner/Le Marin du Nord*, XI, No. 3 (July/juillet 2001), pp. 41-53.

_____ and Nordvik, Helge W. "Coping With the International Shipping Crisis of the 1970s: A Study of Management Responses in Norwegian Oil Tanker Companies." *International Journal of Maritime History*, VIII, No. 2 (1996), pp. 33-69.

Thanopoulou, Helen A. "The Growth of Fleets Registered in the Newly-emerging Maritime Countries and Maritime Crises." *Maritime Policy and Management*, XXII, No. 1 (1995), pp. 51-62.

_____. "What Price the Flag? The Terms of Competitiveness in Shipping." *Marine Policy*, XXII, No. 5 (1998), pp. 359-374.

Theotokas, John. "Organizational and Managerial Patterns of Greek-Owned Shipping Enterprises and the Internationalization Process from the Interwar Period to 1990." In Starkey, David J. and Harlaftis, Gelina (eds.). *Global Markets: The Internationalization of the Sea Transport Industries since 1850*. St. John's: International Maritime Economic History Association, Research in Maritime History No. 14, 1998, pp. 303-318.

Thowsen, Atle. "Skipsfart og planøkonomi. Kontraherings- og lisensieringspolitikken overfor norsk skipsfart i den første etterkrigstiden (1945-1953)" ["Shipping and the Mixed Economy: The Contracting and Licensing Policies with Regard to Norwegian Shipping in the Initial Postwar Period (1945-53)"]. *Sjøfartshistorisk Årbok 1985* [*Norwegian Yearbook of Maritime History 1985*]. Bergen: Bergen Maritime Museum, 1986, pp. 7-36.

_____. "Norsk sjøfartshistorie – periferi eller sentrum i norsk historieforskning" ["Norwegian Maritime History – Periphery or Centre of Norwegian Historical Research"]. *Sjøfartshistorisk Årbok 1972* [*Norwegian Yearbook of Maritime History 1972*]. Bergen: Bergen Maritime Museum, 1973, pp. 9-38.

Tolofari, S.R. "Open Registry Costs and Freight Rates: Are They Related?" *International Journal of Transport Economics*, XIV, No. 1 (1987), pp. 85-103.

_____; Button, K.J.; and Pitfield, D.E. "An Econometric Analysis of the Cost Structure of the Tank Sector of the Shipping Industry." *International Journal of Transport Economics*, XIV, No. 1 (1987), pp. 71-84.

Tronsmo, Per. *Omstilling og organisasjonskultur: lederskap og overlevelse-sevne i tre norske rederier* [*Transformation and Organisational Culture: Management and the Ability to Survive in Three Norwegian Shipowning Companies*]. Oslo: Bedriftsøkonomens Forlag, 1987.

Tvedt, Jostein. "Market Structure, Freight Rates and Assets in Bulk Shipping." Unpublished Dr. oecon. thesis, Norwegian School of Economics and Business Administration, 1995.

United States. Department of Commerce. *Maritime Subsidies 1978*. Washington, DC: United States Government Printing Office, 1978.

Van der Wee, Herman. *Prosperity and Upheaval. The World Economy 1945-1980*. London: Viking, 1986.

Vergottis, Andreas Rokos. "The City University Econometric Model of the Shipping Markets." In Strandenes, Siri Pettersen; Svendsen, Arnljot Strømme; and Wergeland, Tor (eds.). *Shipping Strategies and Bulk Shipping in the 1990s*. Bergen: Centre for International Business, Norwegian School of Economics and Business Administration, 1989, pp. 24-38.

Vikøren, David. "Den internasjonale situasjon for norsk skipsfart" ["The International Situation for Norwegian Shipping"]. *Statsøkonomisk Tidsskrift*, LXXXV, No. 1 (1971), pp. 32-47.

Ville, Simon P. and Williams, David M. (eds.). *Management, Finance and Industrial Relations in Maritime Industries: Essays in International Maritime and Business History*. St. John's: International Maritime Economic History Association, Research in Maritime History No. 6, 1994.

Walker, Fred M. (ed.). *European Shipbuilding: One Hundred Years of Change*. London: Marine Publications International, 1983.

Wergeland, Tor. *Et konkurransedyktig Norge - Norsk skipsfarts konkurranseevne* [*A Competitive Norway – The Competitiveness of Norwegian Shipping*]. Bergen: Foundation for Research in Economics and Business Administration, Norwegian School of Economics and Business Administration, 1992.

Wijnolst, Niko and Wergeland, Tor. *Shipping*. Delft: Delft University Press, 1996.

World Trade Organisation. *International Trade Statistics 2005*. Geneva: World Trade Organisation, 2005.

Yeats, Alexander J. *Shipping and Development Policy: An Integrated Assessment*. New York: Praeger Publishers, 1981.

Yolland, John B. "Ship Finance and Euro-markets." *Maritime Policy and Management*, VI, No. 3 (1979), pp. 175-181.

Zannetos, Zenon S. *The Theory of Oil Tankship Rates*. Cambridge, MA: MIT Press, 1966.

_____. "Persistent Economic Misconceptions in the Transportation of Oil by Sea." *Maritime Studies and Management*, I (1973), pp. 107-118.

_____. "Oil Tanker Markets: Continuity amidst Change." In Gordon, Richard L.; Jacoby, Henry D.; and Zimmerman, Martin B. (eds.). *Energy: Markets and Regulation: Essays in Honor of M.A. Adelman*. Cambridge, MA: MIT Press, 1987, pp. 235-257.

_____. "Market and Cost Structure in Shipping." In Lorange, Peter and Norman, Victor D. (eds.). *Shipping Management*. Bergen: AS John Grieg, 1973, pp. 35-46.

Økonomisk Rapport. Various issues, 1970-1980.

Østensjø, Pernille. *Chemical Shipping*. Bergen: Foundation for Research in Economics and Business Administration, Norwegian School of Economics and Business Administration, 1992.

Printed and bound by CPI Group (UK) Ltd, Croydon, CR0 4YY

27/10/2024

14580409-0002